Contents

Preface

Acknowledgements

Foreword

1	Rethinking construction	8
	Early demands for change	9
	NEC and good project management	9
	Partnering	10
	New demands for change	15
2	How NEC is intended to be used	17
	What NEC provides	18
	Changing attitudes	19
	Drafting style	19
	NEC as the reference document	20
	Decision to use NEC	20
	Selecting the project manager	22
	What the project manager does	23
	What the contractor does	24
	How NEC deals with risk	26
	How NEC deals with quality	26
	How NEC deals with time	27
	How NEC deals with money	30
	How NEC deals with changes	34
	How NEC deals with disputes	37
	Implementing NEC throughout an organisation	38
3	Case studies of NEC in use	41
	ABSA Bank Towers in Johannesburg	42
	GDG Management	48
4	Mature partnering	54
	Teamworking and networks	55
	The NEC Partnering Option	59
	Individual contracts	62
	Individually designed design build	64
	Individually designed construction management	65
	Projects using well-developed designs	66
5	Case studies of NEC and partnering	71
	The Environment Agency's use of NEC as a partnering tool	72
	Costain Limited's experience of NEC and partnering	84
	The Halcrow Group's approach to NEC and partnering	95
6	The NEC Partering Option	110
	Introduction to the NEC Partnering Option	111
	The Conditions: Option X12: Partnering	112
	Commentary on the Clauses	114
	Methodology for setting up contracts using Option X12: Partnering	119
	Option X12: Partnering	119
	The methodology for ECC and PSC	124
	The methodology when using an ECSC	125
	Appendix: Option X12 the NEC Partering Option	126

About the authors

Professor John Bennett DSc, FRICS

John Bennett is Professor of Quantity Surveying in the Department of Construction Management & Engineering at The University of Reading. He was Director of the Centre for Strategic Studies in Construction from 1986 to 1997, where he undertook extensive research into partnering and project management. He is the principal author of over 50 reports and books, including Building Britain 2001, International Construction Project Management, Trusting the Team, Designing and Building a World Class Industry, The Seven Pillars of Partnering and Construction – The Third Way.

John Bennett was employed as Professor at the University of Tokyo during 1991, researching the management methods used by the big Japanese contractors. He was Chairman of the SMM Development Unit that drafted SMM7, and the first Chairman of the joint ACE/BEC/RIBA/RICS Building Project Information Committee. He was the principal academic member of the consortium, led by WS Atkins International, which produced the Strategic Study of the Construction Sectors for the Commission of the European Union that provides the basis for the Commission's strategy towards construction.

He was consulted by the teams which produced the Latham and Egan Reports and his research and ideas influenced their recommendations.

Andrew Baird BSc, CEng, FICE, M(SA)ICE, MAPM

Andrew Baird graduated as a civil engineer in the UK in 1963. He has held senior positions on major projects for consulting engineers, contractors and, since 1982, for ESKOM, the South African electricity utility, as their Corporate Contracts Consultant. In 1998 he started his own firm, Engineering Contract Strategies, specialising in professional services on contracts.

He was invited to join the New Engineering Contract Panel prior to publication of the first edition of NEC, and has been closely involved with the subsequent development of the complete family of documents. Although Andrew Baird is a member of the NEC Panel, the contribution that he has made to this Guide is entirely his own.

In 1993 he led the introduction of NEC into ESKOM at a time when they were undertaking a massive expansion programme involving the engineering and construction of several major generating plants. Since 1998 he has consulted to UK and South African Government agencies, clients, professional project managers, contractors, consultants, banks and law firms on the application of NEC in all sectors of the engineering and construction industries. He lectures on NEC in many countries and has published numerous papers on the NEC system. As a result, Andrew Baird probably has more direct experience than anyone of using NEC across the full procurement spectrum of works, services and supply.

Note

The comment and views expressed by both authors do not necessarily represent the views of the NEC Panel, Thomas Telford or the Institution of Civil Engineers.

Preface

About The Guide

The immediate trigger for The Guide was a conference in Johannesburg during August 1999 on the Practical Applications of Partnering using the New Engineering Contract (NEC) family of contracts. This was organised by the South African NEC Users Group Association in conjunction with Professor John Bennett. Over two days the conference identified the outline of the changes in culture, systems and behaviour required to move away from traditional practice, through the use of NEC, to mature partnering. At the time of the Johannesburg conference NEC included clauses requiring the parties to adopt partnering attitudes. However, the integration of the separate contracts between a customer and all the consultants and contractors required for a project was more implicit than explicit. This has now been changed by the publication in 2001 of the NEC Partnering Option.

The Guide builds on all that background work to provide practical guidance on effective ways of using NEC, including the NEC Partnering Option, for all sectors of the engineering and construction industries and their customers. The Guide does not replace any NEC contracts or Guidance Notes, which should always be consulted for a more detailed statement of how each contract works.

The Guide describes in Chapter 1 the key steps in the engineering and construction industry's adoption of NEC and partnering. Then Chapter 2 describes how NEC works in support of best practice project management. The two case studies in Chapter 3 demonstrate how effective NEC is when it is used as it was intended. Chapter 4 describes how best practice partnering depends on competent teams and well-developed project management. Then it explains how The NEC Partnering Option provides a clear framework for partnering that builds on NEC's established strengths. Chapter 5 describes how this is already a practical reality in three case studies of teams already using NEC as the basis for partnering in ways that provide significant benefits for the customers, consultants and contractors involved.

The Guide recognises that the NEC family of contracts and guidance notes provides an integrated approach. This is referred to throughout the report as NEC. Where a more specific reference is intended, the particular document is named using the following abbreviations:

- ECC means the NEC Engineering and Construction Contract, 2nd Edition.

- ECSC means the NEC Engineering and Construction Short Contract, 1st Edition

- ECS means the NEC Engineering and Construction Subcontract, 2nd Edition

- PSC means the NEC Professional Services Contract, 2nd Edition

- AC means the NEC Adjudicator's Contract, 2nd Edition

Acknowledgements

The authors gratefully acknowledge the generous contributions of time made by the many people who helped produce the case studies described in The Guide. It has been a particular help to have the enthusiastic support of the New Engineering Contract Panel, and much help and advice from Thomas Telford's staff.

Foreword

This Guide to using NEC and partnering marks an important stage in the growing influence of NEC. Professor John Bennett's extensive research into partnering and Andrew Baird's long experience of using NEC combine to provide an authoritative guide that explains in practical terms how NEC-based project management and partnering provides the best approach for all types of engineering and construction projects.

NEC's central role in developing best practice is illustrated by the case studies which show how effectively NEC supports project management and provides a secure basis for partnering. The benefits to customers, consultants and contractors described in the case studies already achieve the targets set by the Latham and Egan Reports, and so confirm that NEC is a crucial tool in enabling engineering and construction firms to meet the increasingly tough demands made by leading customers.

NEC's substantial achievements now take an important step forward with the publication of The NEC Partnering Option, which provides a clear contractual framework for best practice partnering. The Guide explains how it can be used to further extend the benefits of using NEC.

We are therefore delighted to have this opportunity of urging everyone involved in engineering and construction projects – customers, consultants and contractors – to read The Guide and to put its advice into practice.

PETER HIGGINS

Chairman, NEC Panel

GRAHAM CLARKSON

Chairman, NEC Users Group

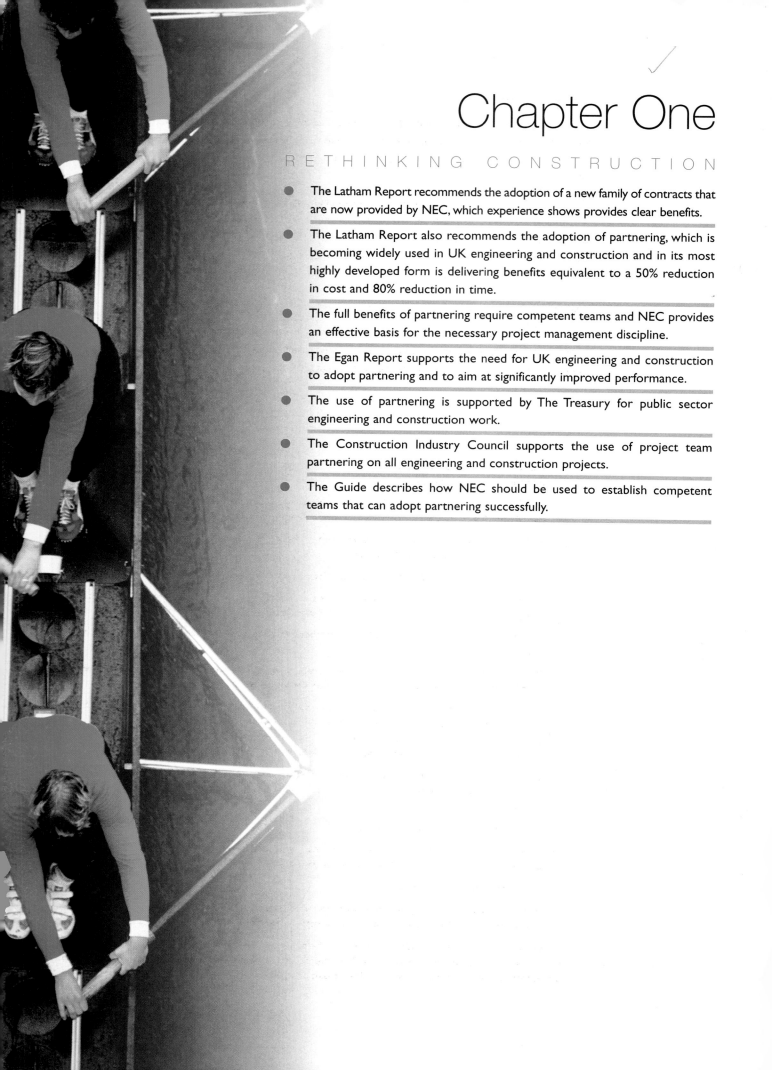

Chapter One

- The Latham Report recommends the adoption of a new family of contracts that are now provided by NEC, which experience shows provides clear benefits.

- The Latham Report also recommends the adoption of partnering, which is becoming widely used in UK engineering and construction and in its most highly developed form is delivering benefits equivalent to a 50% reduction in cost and 80% reduction in time.

- The full benefits of partnering require competent teams and NEC provides an effective basis for the necessary project management discipline.

- The Egan Report supports the need for UK engineering and construction to adopt partnering and to aim at significantly improved performance.

- The use of partnering is supported by The Treasury for public sector engineering and construction work.

- The Construction Industry Council supports the use of project team partnering on all engineering and construction projects.

- The Guide describes how NEC should be used to establish competent teams that can adopt partnering successfully.

Early demands for change

The UK engineering and construction industries have the advantage of two widely supported reports, the Latham and Egan Reports, describing how and why practice needs to change. Although the two reports are both recent and have broadly the same objectives of improving the industry's performance, their recommendations are very different.

The Latham Report, published in 1994, describes the industry's adversarial attitudes and the resulting weak performance that provides customers with little assurance that the costs, times or quality standards stated in the contract will be achieved. It then makes detailed recommendations to help the industry improve its traditional project-based approach. Central to this is the adoption of a new family of contracts based on modern conditions of contract.

NEC and good project management

NEC goes further than this in seeking to encourage good project management. This begins with the employer appointing a project manager to manage the project on his behalf. Many traditional forms of contract create conflicts of interest for the person charged with representing the customer's interests. ECC

The Latham Report's criteria for modern conditions of contract

1   A specific duty for all parties to deal fairly with each other, and with their subcontractors, specialists and suppliers, in an atmosphere of mutual co-operation.

2   Firm duties of teamwork, with shared financial motivation to pursue those objectives. These should involve a general presumption to achieve 'win-win' solutions to problems that may arise during the course of the project.

3   A wholly interrelated package of documents which clearly defines the roles and duties of all involved, and which is suitable for all types of project and for any procurement route.

4   Easily comprehensible language and with Guidance Notes attached.

5   Separation of the role of contract administrator, project or lead manager and adjudicator. The project or lead manager should be clearly defined as the customer's representative.

6   A choice of allocations of risks, to be decided as appropriate to each project but then allocated to the party best able to manage, estimate and carry the risk.

7   Taking all reasonable steps to avoid changes to pre-planned works information. But, where variations do occur, they should be priced in advance, with provision for independent adjudication if agreement cannot be reached.

8   Express provision for assessing interim payments by methods other than monthly valuations, i.e. milestones, activity schedules or payment schedules. Such arrangements must also be reflected in the related subcontract documentation. The eventual aim should be to phase out the traditional system of monthly measurement or re-measurement, but meanwhile provision should be made for it.

9   Clearly setting out the period within which interim payments must be made to all participants in the process, failing which they will have automatic right to compensation, involving payment of interest at a sufficiently heavy rate to deter slow payment.

10   Providing for secure trust fund routes of payment.

11   While taking all possible steps to avoid conflict on site, providing for speedy dispute resolution if any conflict arises, by a pre-determined impartial adjudicator, referee or expert.

12   Providing incentives for exceptional performance.

13   Making provision where appropriate for advance mobilisation payments (if necessary, bonded) to contractors and subcontractors, including in respect of off-site prefabricated materials provided by part of the construction team.

The Latham Report concludes that NEC contains virtually all these requirements and lists the few changes needed to bring it into line with these modern conditions of contract. The changes have since been incorporated, and so NEC now fully provides the modern conditions of contract recommended by the Latham Report.

avoids this by being absolutely clear that the project manager's responsibility is to look after the customer's interests and, by requiring the appointment of a supervisor, to ensure that quality standards are met, and an independent adjudicator to decide upon any disputes.

The project manager begins by managing the production of a clear, detailed and complete description of the required work to be carried out by the contractor. ECC does not mention designers, and all design information provided to the contractor is channelled through the project manager. A detailed programme for the construction work is required to be produced once the contractor has been appointed, usually by competitive tender. This must include method statements and the planned resource use in sufficient detail to allow the costs and times for each activity to be understood and accepted by the project manager. Construction then takes place in accordance with this agreed plan.

The NEC approach accepts that changes may be required, but maintains the principles of good management by requiring the effects on the plan to be considered and agreed before any change is made. This is achieved by requiring the project manager and contractor to notify each other of any matters that may alter the plan or affect the performance of the new facility in use. These early warnings give both parties the right to call a meeting to decide how to deal with the matter. Once it becomes apparent that the costs or times in the agreed plan should be changed, the project manager or contractor notifies the other of a compensation event. This triggers a procedure designed to ensure that the change is evaluated quickly in a manner that all parties can regard as fair.

The overall effect is that the project manager and contractor share an up-to-date and agreed record of the work that is to be done, and its planned costs and times. This encourages far more open communication than is normal in traditional practice and so helps contractors concentrate on working efficiently and delivering the quality and time performance required by their customers.

Achieving these benefits does involve a considerable amount of paperwork. Nevertheless, NEC's encouragement of a formal approach to project management reflects an important principle of best practice, which is that improving performance depends on first establishing control. Without control, initiatives and innovations have unpredictable results and often make little improvement, or even result in worse performance. However, once processes are under control, new ideas can be introduced with the confidence that they are likely to deliver benefits.

There is little doubt that NEC provides just this kind of control and so must be seen as an important step forward from the industry's traditional methods. This is confirmed by the case studies in Chapter 3, which consistently show that using NEC does, as Latham predicted, reduce adversarial attitudes and give customers greater assurance that their new facilities will be delivered at the cost, time and quality agreed in the contract. NEC's approach to project management is an important factor in achieving these improvements. It does, however, set limits to the practical application of one of the Latham Report's other recommendations.

Partnering

The Latham Report recommends that the industry and its customers use partnering. This recommendation is couched in tentative terms because at the time the report was written there was very little experience of using partnering in UK engineering or construction. This situation has now changed dramatically, as described in Trusting the Team and The Seven Pillars of Partnering. In his foreword to Trusting the Team, Sir Michael Latham says that the report goes to the heart of the problems facing the industry and its customers by providing clear best practice guidance aimed at replacing adversarial attitudes with a new approach based on practical actions which encourage co-operative teamworking. He concludes by confirming his view that the industry and its customers need to embark upon partnering if the Latham Report's aims are to be met.

Partnering first emerged in the car industry in the USA as a response to competition from Japan's more efficient methods of manufacturing. These methods deliberately develop co-operative, long-term relationships between all the separate firms involved in designing, manufacturing and assembling cars. They also rely on technically competent teams not only making design and engineering decisions, but also planning, measuring and improving their own processes and deciding which other teams, suppliers and customers they should interact with in order to undertake their work. Team decisions are based on values and strategies that are debated and negotiated throughout the resulting multi-firm organisations. The outcomes are levels of efficiency, innovation and quality which traditional manufacturing methods are unable to match.

Partnering helped the car industry, first in the USA and then worldwide, to improve and so challenge Japan's dominant position. All major car producers now use what are essentially Japanese methods based on partnering. This success has inevitably influenced other industries, including construction. Partnering has been applied to many engineering and construction projects in the USA since the mid-1980s. It first appeared in UK engineering and construction in the North Sea oil and gas industries. At the time Trusting the Team was written this was still in its early stages, and partnering had not spread to the building and civil engineering sectors.

As a result, Trusting the Team is based on research into Japanese construction practice, which uses methods based on Japanese manufacturing practice, case studies of partnering in engineering and construction in the USA and case studies of informal long-term relationships in UK engineering and construction. It is significant that, on this basis, Trusting the Team reflects essentially the same principles of good project management as are now embodied in NEC. They are expressed as: agreeing mutual objectives which take account of the interests of the customer, consultants, contractor and specialists; deciding at the start of projects how problems will be resolved; and aiming at continuous improvements in performance from project to project.

This approach has, as Sir Michael Latham hoped, encouraged the use of partnering so that the subsequent report, The Seven Pillars of Partnering, published in 1998, is based on case studies of the practical application of partnering in UK engineering and construction. It describes how partnering has been adopted by leading customers of the industry who have worked with groups of carefully selected consultants and contractors to develop and refine the approach. As a result, The Seven Pillars of Partnering describes three successive generations of partnering as firms move away from traditional methods and build up their experience of working co-operatively over several projects. Many case studies are described to illustrate how leading UK engineering and construction firms have discovered the benefits of becoming fully committed to partnering in all their relationships with other organisations. The report also includes data that measure the cost and time benefits provided by each generation of partnering.

First-generation partnering

First-generation partnering is essentially the Trusting the Team model applied on a project-by-project basis. Practice has served to redefine the three essential elements of partnering as mutual objectives, decision systems and continuous improvements. The original element of problem resolution was found to be too narrow and negative and has been extended into the principle that project teams should agree the decision system they intend to use. The main choice they need to make in doing this is whether they can use an existing, established approach or whether they need to create a new answer. This choice crucially influences the nature of the work they will need to do and therefore the kind of decision system they should put in place.

NEC encourages the three essential elements of first-generation partnering. Enforcing better quality contract documentation makes the basis of the contract clearer, and so makes it easier to find mutual objectives. NEC is designed as a formal decision system of actions to be taken within specified time periods which helps people concentrate on productive work. This is in sharp contrast to traditional contracts, which stress rights and obligations when things go wrong. NEC's range of options allows the decision system to be matched to the specific project. The third element, continuous improvement, is encouraged by NEC's early warning clauses and provisions for considering alternative ways of dealing with compensation events.

The Seven Pillars of Partnering shows that when project teams put the three essential elements in place they achieve substantial improvements in performance. Depending on the particular objectives agreed by the project team, costs can be 30% and sometimes 40% lower than for similar traditional projects.

Second-generation partnering

The Seven Pillars of Partnering mainly describes second-generation partnering, which achieves even bigger benefits by explicitly adding a long-term strategy to a series of projects undertaken for one customer. The seven pillars are each a set of management actions that put this strategic partnering into effect and at their best achieve cost reductions of up to 40% and time reductions of up to 50% compared with traditional methods.

These important improvements in performance result from the industry moving beyond its traditional view that engineering and construction projects should be treated as one-off and individual. By taking a long-term view over a series of projects, it also moves beyond established ideas about good project management. In second-generation partnering, the management of individual projects is largely predetermined outside of the project, so that project teams work within a strategy developed by a strategic team. This is further reinforced by keeping project teams together so that efficient ways of working become instinctive and few management decisions are needed. Task forces set up by the strategic team outside of individual projects develop new ideas and innovations. When they are ready to be tested on a live project, the extent and timing of their application is determined by the strategic team. So project work becomes much more routine and efficient than with traditional approaches.

Second-generation partnering is dependent on a customer with a regular programme of engineering or construction projects deciding to partner with a group of industry firms. While customers with regular programmes account for much of the industry's work, many projects have individual, one-off customers. Some firms, recognising the benefits of continuity provided by second-generation partnering, are beginning to take initiatives aimed at providing continuity for themselves, even when much of their work is individual projects. The Seven Pillars of Partnering describes this as a third generation of partnering.

The Seven Pillars of Partnering

Strategy

Second-generation partnering arrangements create an organisation formed of several firms. Their joint work is guided by explicit strategies devised by strategic teams comprising senior managers from all the firms involved. They deal with issues that go beyond individual projects by aiming at specific improvements in performance over a series of projects. In this way they establish long-term objectives that take account of the interests of the customer and all the engineering and construction firms involved.

The strategy is put into effect by the strategic team establishing best practice, and setting targets for project teams that embody specific measurable improvements. Strategic teams often base these targets on the work of task forces set up to develop better answers to specific aspects of the organisation's work. Task forces may look for ways of reducing costs, reducing construction times, achieving zero defects, improving safety or any other major improvement. The results of their work are initially put into effect on a project chosen by the strategic team. Then, if the new idea works well, it becomes part of the organisation's established best practice.

Membership

Second-generation partnering brings together firms that can make a significant contribution towards improving the performance of the facilities and services provided for a customer with a regular programme of engineering and construction work. This will normally include the customer, main designers, main contractor and key specialist contractors. The firms should be selected carefully to ensure they can provide people experienced in their particular discipline who are able to develop new and better answers as part of a multi-disciplinary team.

One important aim of the membership pillar should be to appoint all the key members of project teams at the start of individual projects: this means everyone who is able to make a significant difference to the success of the outcome. Thus it may include, from the customer's organisation, representatives of the people who will use the new facility, the finance department, facilities managers and a project manager. It may include representatives of neighbours and special interest groups affected by the proposed work. It will almost certainly include the main designers, contractors, cost managers and specialist contractors. As far as possible the people appointed should be used to working together and have considerable experience of the kind of work required.

Equity

An important aim of the equity pillar is to enable project teams to concentrate on making the project a success for everyone involved by making it unnecessary for them to worry about individual financial matters. The key principles are that the price to the customer should be based on the business case for the new facility. That is the value to the customer, and so should be accepted as the overall budget. The construction firms involved should all be guaranteed a fair profit and reimbursement for their fixed overheads. Then the project team's aim is to produce the best possible facility for no more than the budget, less the guaranteed profits and fixed overheads. Best practice achieves this by reimbursing the firms involved all their direct costs based on open book accounts, tough audits to ensure that costs are correctly accounted for and rigorous cost control.

Strategic teams need to agree how savings or cost over-runs are to be dealt with. This should be part of the strategic arrangement so that disputes over money are not allowed to divert project teams away from their search for the best possible answers within the overall budget. Another issue for the strategic team is how long-term development work, usually undertaken by task forces, is to be funded.

Within this overall approach the broad aims are for the facilities and services provided for the customer to represent better and better value; this usually means, amongst other improvements, that prices are lower. At the same time the construction firms should get higher and more reliable profits.

Integration

The firms involved in a second-generation partnering arrangement need to integrate their standards, procedures, methods and cultures in order to achieve the continuous improvements in performance that are the hallmark of successful partnering. This inevitably takes time and steady, consistent effort from the strategic team. It often requires training in such things as co-operative behaviour, quality, time and cost control procedures, measuring and analysing processes, and creative techniques to help the search for better answers. It often requires the firms to alter their internal procedures and spend time explaining the benefits of partnering, e.g. to the legal, buying and finance departments. They may be wedded to competitive tendering and tough contracts, and so need to be persuaded of the benefits of partnering with firms they have previously enjoyed taking advantage of whenever possible. The aim of the integration pillar is to form a virtual organisation in which people drawn from separate firms work as integrated teams.

Benchmarks

The primary aim of partnering is to improve performance. It is impossible to be sure that this is being achieved sufficiently well unless performance is measured in terms which allow individual projects to be compared. The resulting benchmarks should reflect the strategic objectives and be expressed in terms that the members of the strategic team can understand and believe.

Benchmarks should allow the partnering organisation's projects to be compared to allow targets to be set and to ensure that judgements about the overall performance can be soundly based. They should also allow comparisons with similar facilities and services being provided for other customers. There is always a temptation for senior managers to test the market. A rogue bid can all too easily cause good work and well-established relationships to be thrown away. This is why it is essential for partnering organisations to have benchmarks which demonstrate that they are competitive with what is available elsewhere in the market.

Benchmarks can be used in benchmarking which is a structured approach to comparing the organisation's performance with the best in the world and using the results to guide a search for better answers.

Project processes

One of the most distinctive features of second-generation partnering is that it systematically develops the processes used by project teams. This is directed by strategic teams using ideas developed by task forces and feedback about good ideas arising on individual projects. The resulting best practice is embodied in the standards and procedures used by project teams.

Appropriate standards and procedures are developed by second-generation partnering organisations for each different type of project required by the customer. These may include projects requiring individual designs just as much as those able to use as well-developed designs. The common aim is for every project team to work with a well-developed approach which they fully understand. The results are high levels of performance that help realise the organisation's long-term strategy.

Feedback

The seven pillars form a controlled system which delivers continuous improvements in performance guided by feedback, which provides the seventh pillar. Feedback is essential for control and for improvement. Individual projects need feedback on their performance so they meet their quality, time and cost targets. Long-term improvement requires feedback to flow from project to project so that good ideas are captured, studied, developed, tested and incorporated in standards and procedures.

Third-generation partnering

Third-generation partnering provides a mature approach in which the engineering and construction industry has adopted practices normally used in manufacturing industries. This is achieved by a group of partnering firms taking responsibility for the complete cycle of use-development-production.

Firms involved in third-generation partnering develop and market products and services aimed at categories of customers. As in other manufacturing industries, this provides continuity, which in construction is already allowing very efficient methods to develop that can give cost reductions of up to 50% and time reductions of up to 80% compared with traditional methods. The role of individual projects is determined by the nature of the work. Thus for complicated work there remains a need for traditional project management, even though it works within a tightly defined framework of standards and procedures. However, for much of the industry's mainstream work, one-off projects scarcely exist and new facilities result from continuous on-going processes.

Key features of third-generation partnering

Overall aims

Mature engineering and construction industries aim at sustainable development that serves sustainable communities. They think and act long-term by aiming at continuous improvements in performance that both justify and deliver the high levels of profits needed to invest in innovation.

Outputs

Mature engineering and construction industries use marketing to identify specific categories of customers and then produce products and services aimed at their needs. This provides two broad categories of facilities and supporting services. First, there is mainstream work providing standardised products and well-developed packages of services. Second, there is new-stream work providing new designs and innovative technologies in original ways. Both categories are developed as a continuing process so that the industry progressively provides better and better value for all its customers.

Organisation

Mature engineering and construction industries organise themselves into groups of firms that work together in long-term relationships. They develop integrated systems and procedures and a common culture of trust and co-operation. These organisations invest long-term in improving their performance, products and services. This includes investing in training, research and development, and market research.

Project organisation

Projects are undertaken by well-established project teams closely matched to the type of work needed to meet agreed targets and constraints. They work by identifying the standards and procedures that apply and putting them into practice on an almost automatic basis so they can concentrate on non-standard elements and looking for improvements.

Information

Information about performance and best practice flows freely throughout the industry, so there are well-founded benchmarks to guide decisions by customers and industry firms. The benchmarks provide objective measures of the key interests of all stakeholders in terms that relate to world class performance. It is widely understood that secrecy is damaging and the industry has sufficient confidence to be open about all aspects of its work.

Systems

The industry works through standards and procedures based on the principles of integrated systems controlled on the basis of feedback. All parts of the industry work steadily at continuously improving their standards and procedures but are vigilant in identifying the need for a radical step change in response to market or technological changes.

Teamworking

It is implicit in mature engineering and construction industries that responsibilities rest with teams. It is understood that problems nearly always come from weaknesses in systems and so the industry has a no-blame culture. Problems and mistakes are seen as opportunities to find better answers. A key feature of mature practice is that teams are kept together so they can work at becoming ever more effective. In practice this has to be balanced against the need to provide career paths for individuals. So there is dynamism within relatively stable team structures.

Liabilities

Individual liabilities destroy teamworking and they do not motivate people to achieve their best work. So liabilities are carried either by the customer who wants to develop new ideas and takes the incentive in directing construction firms, or by a substantial construction organisation that is very likely to still be in business in 10 or 20 years time. The substantial construction organisation may be a single large firm or a group of firms with a long-term basis for working together. In the early days of such arrangements, insurance plays a key part in providing customers with the confidence to invest in new facilities. In the long-term the industry needs to be able to convince customers to rely on their reputation, ideally embodied in a brand name.

New demands for change

This mature approach provides an important part of the background against which the Egan Report was produced for the UK's new Labour government. Published in 1998, it goes a great deal further than the Latham Report's recommendations and so inevitably goes beyond the approach embodied in NEC.

Key recommendations in the Egan Report

1. The industry and its major customers need to rethink construction so as to match the performance of the best consumer-led manufacturing and service industries.

2. The industry should organise its work so that it offers customers brand-named products and services which they can trust to provide reliably good value.

3. The industry should work through long-term relationships using partnering, which aims at continuous improvements in performance.

4. Benefits from improved performance should be shared in an openly fair basis so that everyone involved has a real motivation to search for better answers.

5. Project teams should include design, manufacturing and construction skills from day one so that all aspects of the process are properly considered.

6. Decisions should be guided by feedback from the experience of completed projects so that the industry is able to produce new answers that provide ever better value for customers.

7. Standard products should be used in designs wherever possible because they are cheaper and, in the hands of talented designers, can provide buildings and other facilities that are aesthetically exciting.

8. Continuous improvements in performance should be driven by measured targets, because they are more effective than using competitive tenders.

9. The industry should end its reliance on formal conditions of contract, because in soundly based relationships in which the parties recognise their mutual interdependence contracts add significantly to the cost of projects and add no value for the customer.

These radical proposals are being put into effect on many demonstration projects throughout the industry. The important lessons are being captured and published by the Movement for Innovation, to publicise best practice and the benefits it provides. A significant indication of the changes already underway in the industry is that over 95% of the demonstration projects are using partnering.

Key features of HM Treasury's teamworking, partnering and incentives

1. Projects must aim at achieving value for money.

2. Fees and prices should be established by competition.

3. Success depends on competent people searching for improvements, listening, learning and actively developing their own abilities.

4. Proper and appropriate management structures and procedures are essential.

5. Teamworking should be a core feature of all elements of projects and it should be fostered at team-building workshops.

6. Partnering should be used on all projects and be fully and visibly supported by very senior management from each organisation.

7. Teams should actively seek to reduce the need for paperwork, monitoring other peoples' work, claims, litigation, arbitration and dispute resolution.

8. Incentives should be used to encourage consultants and contractors to provide benefits to customers significantly beyond those contracted for.

9. Risk analysis and management should be explicit.

10. Performance in terms of improvements in value for money must be measured regularly.

11. Successful project teams should be rewarded with continuity of work.

In response to the Egan Report, HM Treasury has recommended that teamworking should be a core requirement and that partnering be adopted as far as possible on all Government contracts.

More generally, the Egan Report gave rise to the Movement for Innovation and a pool of demonstration projects, which provide practical examples of good practice. These important developments have been reinforced by the Construction Industry Council publishing a guide to project team partnering.

An important feature of The Guide is that it explains why it is an advantage to have a multi-party project partnering contract. The NEC Partnering Option

Key features of A Guide to Project Team Partnering

1   The benefits of partnering go beyond saving money to provide greater customer and user satisfaction and more effective ways of working for consultants and contractors.

2   Partnering should focus on the needs of customers and users, which are developed into a well-considered brief, budget and programme with the help of a partnering advisor.

3   Partners should be selected and their fees agreed by means of a systematic quality-based selection process using a carefully considered list of criteria.

4   The most competent project partnering team the customer can afford should be set up very early.

5   Partnering relies on teamworking built up through workshops that encourage everyone to work creatively on a basis of mutual trust and co-operation.

6   Partnering teams need a multi-party project partnering contract that encourages a joint commitment to the project, sets out the common and agreed rules, defines the goals and how to achieve them, states how risks and rewards are managed and how changes are dealt with, provides guidelines for resolving disputes, and lays down how performance is to be measured.

7   The contract should cover collaborative design development, including inputs from consultants, contractors and specialists.

8   Project partnering teams can be guided by a core team of key individuals drawn from project partnering team members.

provides a virtual multi-party contract by incorporating the same Partnering Option into the separate bi-party contracts between the firms involved in a partnering arrangement. This applies whether the arrangement covers just one project or a series of related projects, and it provides the contractual framework for partnering called for by the Construction Industry Council's guide.

Mature partnering depends on teams working with well-developed project management systems. NEC explicitly provides these systems, so the next chapter describes how the NEC family of contracts is intended to be used in providing effective project management.

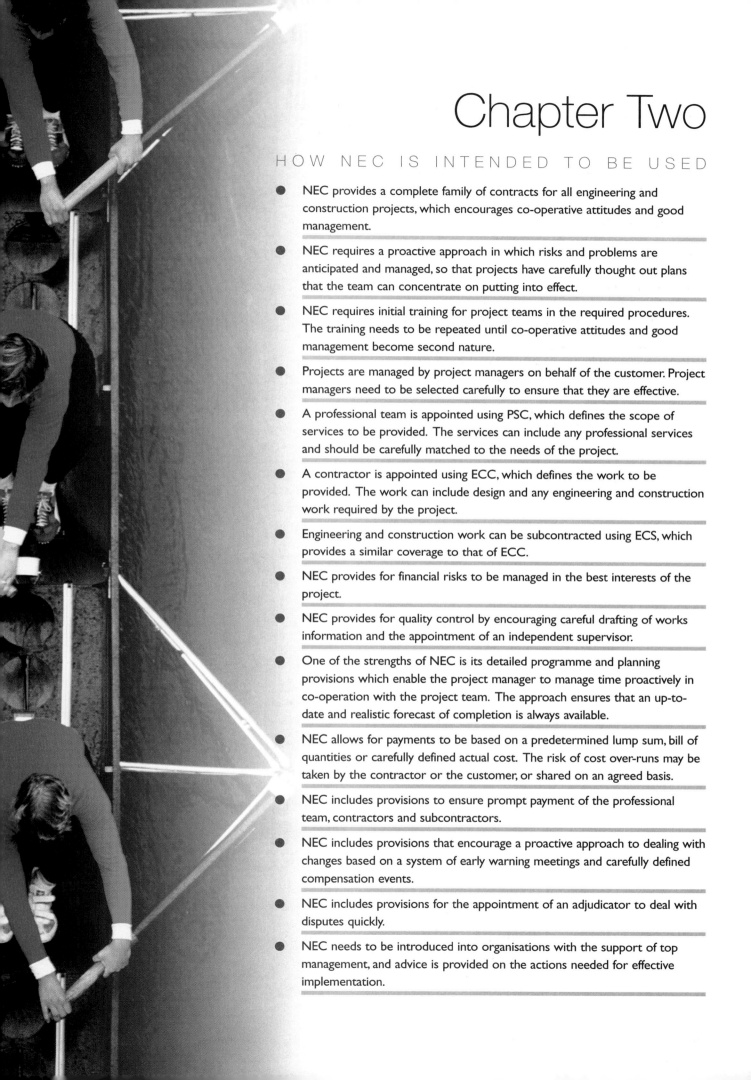

Chapter Two

- NEC provides a complete family of contracts for all engineering and construction projects, which encourages co-operative attitudes and good management.

- NEC requires a proactive approach in which risks and problems are anticipated and managed, so that projects have carefully thought out plans that the team can concentrate on putting into effect.

- NEC requires initial training for project teams in the required procedures. The training needs to be repeated until co-operative attitudes and good management become second nature.

- Projects are managed by project managers on behalf of the customer. Project managers need to be selected carefully to ensure that they are effective.

- A professional team is appointed using PSC, which defines the scope of services to be provided. The services can include any professional services and should be carefully matched to the needs of the project.

- A contractor is appointed using ECC, which defines the work to be provided. The work can include design and any engineering and construction work required by the project.

- Engineering and construction work can be subcontracted using ECS, which provides a similar coverage to that of ECC.

- NEC provides for financial risks to be managed in the best interests of the project.

- NEC provides for quality control by encouraging careful drafting of works information and the appointment of an independent supervisor.

- One of the strengths of NEC is its detailed programme and planning provisions which enable the project manager to manage time proactively in co-operation with the project team. The approach ensures that an up-to-date and realistic forecast of completion is always available.

- NEC allows for payments to be based on a predetermined lump sum, bill of quantities or carefully defined actual cost. The risk of cost over-runs may be taken by the contractor or the customer, or shared on an agreed basis.

- NEC includes provisions to ensure prompt payment of the professional team, contractors and subcontractors.

- NEC includes provisions that encourage a proactive approach to dealing with changes based on a system of early warning meetings and carefully defined compensation events.

- NEC includes provisions for the appointment of an adjudicator to deal with disputes quickly.

- NEC needs to be introduced into organisations with the support of top management, and advice is provided on the actions needed for effective implementation.

What NEC provides

NEC was developed from the outset with the objective of introducing project management into contractual arrangements. Project management is an essential prerequisite to effective partnering. This does not mean tough, adversarial project management, sometimes providing little more than time management. It means a professionally responsible but disciplined approach to teamworking that enables consultants and contractors to do their work in a co-ordinated way, with all participants, including the project manager, taking responsibility for their own actions.

The NEC family of standard contracts is designed to meet the current and future needs of any organisation requiring to procure:

- architectural, engineering and professional services generally;

- design, fabrication and supply of plant; and

- on-site construction and re-construction projects.

This single interlinking system of documents eliminates the need for the plethora of documents used traditionally. Although originally aimed at substantial engineering and building assets, the NEC system now also includes contract forms better suited to lower value contracts.

The first member of the NEC contract family, published in 1993, was designed for use between an employer and a contractor for the design and construction, or construction only, of an engineering or building project. It was known as the New Engineering Contract. As other forms of contract were added to the family, its name was changed to the Engineering and Construction Contract (ECC) and the acronym 'NEC' was reserved for the family as a whole.

Stimulus to good management

Every procedure laid down in NEC contracts has been designed so that its implementation should contribute to, rather than detract from, the effectiveness of management of work required by both parties in any procurement activity. This aspect of the NEC philosophy is founded upon the proposition that:

- foresight applied collaboratively mitigates problems and shrinks risks inherent in engineering and construction work, and

- clear division of function and responsibility helps accountability and motivates people to play their part.

The actions required by the parties to an NEC contract are designed to motivate its users to be solution orientated rather than problem focused. Traditional contracts tend to express only the rights (usually of the employing party) and obligations (usually of the supplying party) and pay little attention to teamwork.

NEC family of documents

The documentation included in each family member is extensive, as demonstrated by that which is available for the ECC.

ECC (the Black Book)
This volume contains all the clauses and schedules comprising the ECC, including:

- core clauses – common to all contracts,

- clauses for each of the main options A to F – one of which should be chosen for a particular contract,

- clauses for each of the secondary options G to Z – each available, if required, for a particular contract,

- Schedule of Cost Components, applicable to main options A to E (providing two methods of computing cost), and

- Contract Data formats parts one and two.

The ECS
This volume contains all the clauses (core and options) and schedules constituting the Engineering and Construction Subcontract, i.e. the form of subcontract that is the equivalent of the complete ECC and which is compatible with it. It is intended to be used for subcontracts to contracts that are let under the ECC.

ECC flow charts
The flow charts show the procedural logic on which the ECC is based. They are available for reference in conjunction with the guidance notes.

ECC guidance notes (the Brown Book)
The purpose of the guidance notes is to explain the background to the ECC, the reasons for some of the provisions and to provide guidance on how to use it.

ECC merged versions
The six merged versions include the clauses for the relevant main option located in their appropriate places amongst the core clauses. Thus the conditions for each main option can be read together. The main option clauses are in bold print for easy identification.

Flexibility

The flexibility provided by the NEC system is demonstrated by the fact that each document within the family provides just about every form of contract likely to be used in that sector of procurement activity.

Responsibilities

Each member of the NEC family expresses the obligations of the party providing the service as being in accordance with:

● the Scope (PSC), or

● the Works Information (ECC, ECS and ECSC).

These simple statements provide much of the NEC's flexibility, as the works information or the scope is developed entirely on a contract-specific basis. This means that if a contractor or consultant is required to undertake a community improvement based upliftment programme as part of his duties within a contract in Africa, the necessary specification can be included in the works information or the scope. Special conditions of contract or adjustments to the standard NEC form are not necessary.

Changing attitudes

The authors of the NEC system were given the directive of finding a better way of managing projects than was then available, in 1987. This was in direct response to clients' general dissatisfaction with the way the industry worked. Amongst the many concerns were the more obvious ones of time and cost over-runs, but others included adversarial business relationships, low profit margins and poor quality.

Any system designed to remedy such concerns must, by definition, give rise to a change of attitude or approach that could, for some participants, be a source of some pain or concern. Changing bad habits is never easy. Introduction of the NEC system has shown this to be the case. The system can be used without a change of mind set or habit. However, the result is likely to be disappointing and may even be disastrous.

Hence the importance of understanding how the NEC is intended to be used, which this chapter and chapter 3 describe, before considering the application of the NEC to mature partnering in chapters 4, 5 and 6.

Drafting style

NEC sets out to encourage better attitudes and better management. This means that some clauses may not be strictly enforceable but, nevertheless, they have a beneficial effect on projects. This tends to worry lawyers, but is seen as helpful by construction professionals who need answers to problems.

In a comparison with traditional contracts, the lack of cross-referencing and use of the present tense (after the first statement of obligation in Clause 10.1) is claimed by some lawyers to give rise to unnecessary uncertainty. However, to the experienced user, the gain in clarity outweighs the need for legal certainty. Some of the provisions, especially for variations, are considered clumsy and overcomplicated to cope easily with the problems that arise on a day-to-day basis on all engineering and construction contracts. The presence of these clauses, however, does encourage a common sense use of short cuts, with the confidence that there is a robust fallback approach if needed.

Using NEC does seem to have kept users out of the legal process as intended, so criticisms have lessened over time. Also, the traditionally legalistic 'yardstick' contract forms, against which NEC has been compared, are themselves now being substantially re-drafted to also make use of language and terminology better understood by industry practitioners, rather than being understood only by lawyers.

With ECS being back to back with ECC, most of what has already been discussed, and that which follows, applies equally to subcontractors as it does to contractors. Readers who mainly act as subcontractors should note that where the term *Employer* or *Project Manager* is used in ECC, in the ECS it is replaced by the term *Contractor*. Likewise, for *Contractor* and Subcontractor in ECC, read *Subcontractor* and Subsubcontractor in ECS.

NEC as the reference document

As is the practice with most standard form contracts, the printed document is not issued with an enquiry for tender, or with the final contract documents. The NEC document, which the client needs to use, along with a statement of the clauses selected from within it, which he needs to apply, is referred to in a document which he prepares, called the Contract Data. The client, or the *Employer* as he is known in NEC terminology, needs to be in possession of an original copy of the chosen NEC printed document before he can use it. In order for contractors and suppliers to tender to, or operate under, an NEC contract they would then purchase a copy of the chosen NEC document for their reference. As more clients use the NEC documents, the same reference document will be re-used for many contracts.

The main options

The main option clauses offer a number of pricing mechanisms or basic allocations of risk between employer and contractor, or consultant, as the case may be. One of the main options must be selected. For example, in the ECC there are six main payment options:

- Options A (activity schedule) and B (bill of quantities) are priced contracts in which the risks of being able to carry out the work at the agreed prices are largely borne by the contractor.

- Options C and D are target contracts in which the financial risks are shared by the employer and the contractor in an agreed proportion.

- Options E and F are two types of cost-reimbursable contract in which the financial risk is largely borne by the employer.

In ECC, irrespective of the main option selected, the boundary between design by employer and design by contractor can be set to suit the chosen strategy. If the works information set down by an employer is only a performance specification, most of the design will be done by the contractor (effectively a 'design and construct' contract). However, if the works information includes detailed drawings and specifications, little design remains for the contractor to complete.

The secondary options

After deciding the main option, the user may choose any of several secondary options. It is not necessary to use any of them.

The Contract Data

In addition to incorporating an NEC contract by reference, the purpose of the Contract Data is also to provide data specific to a particular contract. The Contract Data is the key document in any contract, and even states which legal system will govern the contract.

Decision to use NEC

Using the NEC system requires an objective commitment, which is not generally necessary when using other systems. This costs time and money, which can be seen as an investment, or as insurance, for an improved outcome that in most cases will give rise to lower overall project out-turn cost. It is the client who needs to make the investment, usually in the form of training the team, and a greater concentration on better scope definition at the outset and throughout the progress of a project. The major part of the training involves changing the mind set of the participants. Some clients observe that, as usage of the NEC system spreads, the need for training may reduce. However, it will be a long time before habits are changed and every person in the industry has been exposed to, and bought into, the changed approach. One client, now into his fourth NEC-based project, has decided to insist on training all participants for every project, whether or not the participants are NEC experienced, on the grounds that the training can also be used for scope definition and creative participation before parties become committed to contracts.

The NEC system is dependent on the concept of co-operation. Great emphasis is placed on communications, programming, common sense and the need for clear definition at the outset of various types of information. The system focuses on management and the actions people take, as much as the obligations and liabilities of the parties. If people carry out the actions which the contracts say they are to take, then the obligations of the party they serve will have been met.

NEC contracts are not to be filed away at commencement and dusted down when problems arise. They are as much manuals of project management as they are sets of conditions of contract. They demand constant attention and a high level of administrative support.

In summary, the change of mindset has the following elements:

Traditional mind set	NEC mind set
Allocating risk to one party	Allocating risk to the party best able to manage the effect of the risk
Ignoring the implication of risk	Open debate about creative ways to manage risk
Reacting to problems	Early warning of problems
Recording progress	Planning activities before they are carried out
Re-measurement (usually at the end of the project)	Accurate payment for work done on a regular (usually monthly) basis, with dependable forecasts of the final out-turn cost being available at any stage
Leaving details of a change to later	Forecasting the effects of a change and agreeing on it before it is undertaken
Indifferent reaction to adverse risk events by the party who thinks he may not have to pay for the consequences	Dealing with adverse risk events in a creative way that causes least harm or adverse consequences for all participants
Only correcting defects if instructed	Correcting defects whether notified or not

Reasons for using NEC

Wide experience of using NEC on many hundreds of contracts, worth billions of dollars in all sectors of the engineering and construction industry, has identified the following strategic reasons for using the NEC. It is worthy of note that not one of the serious new users of NEC, to the authors' knowledge, has decided not to use it again. This includes those for whom the first use was not as effective as they expected it to be.

Users of NEC are required to think about how they intend to control capital and other out-sourced expenditure.

Users of traditional contracts often choose them because their professional body recommends that they be used. Traditional contracts are generally designed for application to only one payment strategy, such as a bill of quantities, and to one sector of the engineering and construction industry, e.g. to civil engineering works. There is also a belief that all the commercial controls necessary for a successful outcome will be provided if that contract is used.

When NEC is used, many contract options are available. Making the choice between the main and secondary options available forces the users to seek advice, or at least think far more about the best commercial and contract strategies for the parties to the contract.

NEC provides a single modern commercial platform for the whole engineering and construction industry.

The proliferation of forms of contract has become almost an epidemic. Many of the developers of those forms, whose motives are usually organisation or discipline specific, fail to take account of:

- the harm they are doing to industry relationships,
- an inevitable decrease in overall industry productivity,
- an unnecessary demand on scarce contract management skills, and
- the integration of systems prevalent in the information age.

NEC provides a single integrated contract development and management system based on modern business methods and the techniques and benefits of project management.

Working as a team with the architect, quantity surveyor, and civil, structural, mechanical and electrical engineers all using the same form of contract makes an incredible difference to the process of managing a project.

The traditional approach of using a mix of different discipline-based forms of contract on a single project is almost guaranteed to produce unnecessary divisions and disputes. This is what has led to many clients drafting their own forms of contract. The NEC's multi-discipline approach, with its flexibility to handle design by either party and under contracting strategies that span the whole procurement spectrum, is exactly what clients have been looking for.

NEC enforces a higher quality of contract documentation, a need for effective programming skills, and disciplined contract management by both parties.

Many professional people have recognised that the higher standard of work demanded by the use of NEC is already an obligation of their professional code. However, the adversarial traditional documents currently in use have permitted a lower standard of work with the possibility of being able to shift the blame to others when things go wrong, without it becoming detrimental to relationships with their clients. There is an additional cost to the client for the totally professional approach that the NEC demands, but the improved control of capital expenditure and greater certainty of outcome more than offset this.

Correct application of NEC will ensure that a clients' affairs are managed proactively.

NEC gives the client's advisers a tool to persuade the client of his need to be part of the team and to make the decisions he needs to make within a disciplined time frame. The consequences of the client not doing so are clearly stated in the contract in the form of compensation events. Some clients will still resist such an approach, and NEC is not for them. But there are many who are prepared to see the big picture and the benefits that flow to them from a change of their ways.

When disputes arise, resolving them is far less acrimonious under NEC, and much, much cheaper.

It has been said that there is nowhere to hide when using NEC. This in itself makes dispute resolution much simpler and quicker, and is also a motivating factor to avoid getting into dispute in the first place. Because of the clear role definitions and risk allocation, it is usually easier to identify where things may have gone wrong.

Selecting the project manager

The skill and experience of the project manager (the employer's agent) is critical for any form of contract, but is particularly so for an NEC contract. A sound knowledge of the technology of the project, of the industry, and of resource costing and programming, as well as a will to act as the contract requires, is essential. Once the project manager loses his ability to manage the day-to-day requirements of the contract, by being reactive instead of proactive, the benefits of that certainty of outcome, which the NEC can provide, will be lost.

Selection of the right person to fill the role of project manager is crucial to the success of an NEC contract. It is vital that the project manager be a person who will be proactive and manage the future on a continuous basis. If he reacts to events or becomes a follower he will fail spectacularly. He needs to maintain the respect and support of all the project stakeholders, but particularly of the employer's representative, design professionals, programme and cost specialists, as well as that of the contractor. The key to the role is to manage fairly but firmly.

The use of an in-house project manager generally turns out to be less effective than the use of one from an external professional firm. Corporate matters and staff movements distract from the all-important focus that a project needs. There is also a tendency within large organisations to expect one internal project manager to be responsible for a number of contracts running concurrently. This usually means that the administrative burden forces the project manager to be reactive rather than proactive.

When setting up an ECSC contract, the same attributes are needed in the person chosen to administer the contract on behalf of the client. Most of the ECSC procedures are identical to the equivalent procedure in the ECC. The number of interfaces and other 'players' in an ECSC contract are likely to be fewer and may be less difficult to manage, but it is unwise to rely on this when selecting the person to represent the employer.

What the project manager does

Central to any NEC-based project is the project manager. He is appointed by the employer to undertake the function of managing the project on the employer's behalf. When the project manager is external to the employer's organisation, he will be contracted to the employer under the PSC. The Scope document, which is part of the PSC, will state the duties and responsibilities of the project manager and provide procedures detailing how he is to interact with the employer and other members of the professional team. The project manager can be an individual or a firm, but usually the latter, with a named individual from that firm allocated to the role of *Project Manager* in each construction contract.

The Scope is dependent on the contract strategy chosen for the project. It will be quite different for a project manager required to administer a single Option A ECC contract, from that of a project manager required to undertake the role of construction manager for several direct trade contracts with the employer. This also has implications for the project manager's professional indemnity insurers.

The Scope of the PSC specifies and describes the project manager's duties for the concept, definition, design and tender phase of the project in some detail. However, for the detailed engineering and construction phase it only needs to state that the project manager is required to act in the role of the *Project Manager* under the ECC, stating which main option is to be used. This is because the ECC itself states what the project manager does. It is clear from those actions that the project manager is the agent of the employer and does not perform a quasi-judicial role as well, which is sometimes the case for an Engineer or Architect in traditional construction contract systems. It is intended that the Contract Data for the ECC state the name of a person to act in the role of *Project Manager*. This is because good project management requires a single point of accountability.

Actions of the project manager in ECC are actions of the Contractor under the ECS. The ECSC does not mention the use of an agent but provides for the employer to delegate actions of him under the contract, and it is quite possible that he may delegate to an independent project manager if this is justified by the nature of the work in the contract.

Under the ECC the project manager accepts the contractor's design and programme, and assesses the amount due to the contractor that the employer is required to pay. However, it is not intended that the project manager interfere in any way with the responsibilities and duties of other members of the professional team better qualified to make the required decision or assessment. However, as between the employer and the contractor in ECC, it is the project manager and the supervisor only who communicate on the employer's behalf with the contractor. This is part of the intended objective of providing clarity in the contractual processes. Hence, the project manager will interface with other members of the professional team as required, in order to be able to undertake the actions required of him under the contract. The liabilities of the other members of the professional team are expressed in their respective PSCs with the employer. It is quite likely that one or more of the design professionals will undertake the role of supervisor in the ECC.

There are over 50 actions of the project manager in an Option A ECC contract; many of which are required to be undertaken in short time periods. This requires that the employer delegate the necessary authority to the project manager to enable him to carry out his actions, or has in place a decision-making system that can react to the project manager's requirements in hours rather than days.

The project manager needs to set about his tasks in a very disciplined manner and be supported by good record-keeping and communication systems, as well as having available to him the necessary expertise in the field of programme and cost analysis. He will conduct regular meetings with other members of the professional team. Good practice to date has demonstrated that the project manager needs to hold combined early warning and compensation event meetings at least once a week in order to maintain the time schedules of the ECC clauses.

What the contractor does

The contractor does what the contract says he does. In the ECC there are present-tense verbs such as 'acts', 'notifies' and 'obtains'. To find the express obligations of the contractor it is necessary to search through the contract for present-tense verbs. There are 60 or so applicable to the contractor, the exact number depending on which option has been selected. There are about an equal number applicable to the project manager and the supervisor combined. If these are not done, or not done in accordance with the contract, the contractor has the right to refer the non-action to the adjudicator.

The manner in which the work itself is to be undertaken is described in the works information, along with descriptions of any constraints as to how the work is to be done.

Change of mind set

On the face of it the foregoing is no different to any other contract. However, under the ECC the contractor is encouraged to be an equal member of the employer's team, rather than just a member who undertakes the instructions of others. Quite apart from the fact that the contractor may be doing a proportion of the design, he is completely responsible for planning and programming the work. In the case of Options A and C the contractor will have set down the activities he intends to undertake to complete the work (the work breakdown structure). When it comes to changes, the contractor is required to assess the effect of the change as well as provide alternative methods of undertaking the change.

Because of the nature of the compensation event procedure, the employer's professional team will be motivated to hold design and buildability co-ordination meetings with the contractor before he commences work on site. The changes or additional detail, which arise from these meetings, may constitute compensation events, but resolution at this stage is constructive, with easier identification and control of the cost and time effects, and may well provide extra float for the contractor and cost savings for the employer.

In order to be able to undertake and respond to the actions required of him in the ECC, the contractor will need to allocate his actions to a single person. This person will need to make the same mind-set shift as required of the employer and his professional team. The first change will be to shed the claimsmanship attitude, and substitute it for an early warning attitude by following the procedures of Clause 16. The second major change will be to address the effects of each change, preferably before it occurs, but certainly at the time it occurs, and not leave the consequences to a cumulative end of contract wrangle. Change management will require close liaison with the project manager. A good knowledge of programming, resource monitoring through method statements and cost modelling will be an essential skill for the contractor's representative to have, or to have immediate access to. Careful consideration will need to be given to location of management staff; managing an ECC contract from a location remote from the site is not advisable.

Up and running from the start

All NEC documents include procedures that are designed to be used proactively as well as reactively. None of the procedures need represent any hardship for successful contractors. Dealing with issues quickly will certainly discourage potential disputes. It is acknowledged that contractors may need to persuade the employer to act likewise, especially if the employer has been used to the hands-off approach prevalent in traditional practice. Flooding the project manager with petty early warnings (Clause 16) may not make friends, but it is the procedure that should be invoked to wake up a lazy client and his agents. Many contractors, while making the mind-set change, are reluctant to approach the adjudicator when the project manager has not completed actions he is required to undertake. However, it is the contractor's right to do so, and time constraints for the reference need to be met. Done early on in the contract, and decisively, it is likely to enhance relationships. Left to later, with all the argument as whether the reference is now time barred, destroys relationships.

Time periods for notices, quotations and other communications

Communications, which the contract requires, are generally to be provided within the default period for reply. However, many communications have their own specific time periods. In particular, the contractor is required to notify the employer of a compensation event within two weeks of when he became aware of it, but subcontractors only have one week to similarly notify the contractor. Contractors need to be aware of, and inform their subcontractors about, communication time periods in the main contract with the employer, which are different to those in the ECS.

Records and cost data

Contractor's and subcontractor's cost data and resource records need to be in a format compatible with the ECC, and thus easily applied to compensation event quotations. The data and records should always be available to the project manager for inspection.

Use of standard forms for administration of contract procedures

Administration procedures are well defined in NEC documents, with time periods associated with nearly all actions. This lends itself to the development of standard forms for each of the main communications

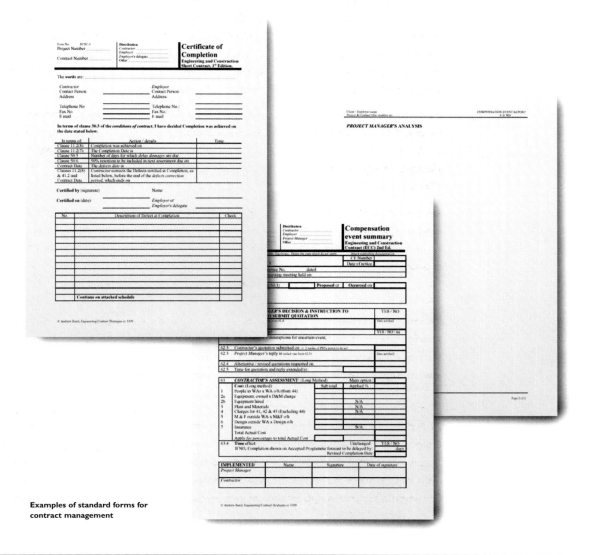

Examples of standard forms for contract management

between project manager, supervisor and contractor. Employers and their project managers will have developed standard forms for their main notice provisions, such as certification for payment and completion. It is expected that contractors will develop forms for the actions that they generally initiate. One professional firm using NEC has developed a complete documentation management system (example documents are shown on page 25).

Quote the relevant clauses

The clauses in NEC documents are drafted as actions of the role players. These actions are further illustrated in the ECC flow charts. Hence, the best way to communicate is to quote the relevant clause number in order to refer to the procedure and avoid subsequent confusion or dispute. Clause 13.7 requires that each communication be kept separate for purposes of clarity. Because of this requirement, minutes of meetings would generally not be held to be communications under the contract unless the meeting was about one specific action in the contract. Even then it is difficult to draft contract instructions in a meeting.

How NEC deals with risk

Limitations of liability to the employer generally, for damage to surrounding property and latent defects, are not addressed fully in the ECC. These matters along with the effects of force majeure need to be addressed in additional conditions of contract in Option Z. They have been more fully addressed in the PSC and ECSC. The approach used in the NEC system is to allocate this type of risk in such a way that when a specified limit has been reached, the effects on the contractor's costs and time for completion above that limit become compensation events. In the ECC, employer's risks relating to physical loss and damage and certain legal liabilities are listed in Section 8 of the core clauses. The list can be added to in Contract Data. If such an employer's risk occurs, it becomes one of the listed compensation events. Allocation and limitation of liability and indemnities generally, will need to be one of the first tasks addressed in a project involving the NEC Partnering Option.

NEC raises an awareness of financial risk allocation in the place of risk avoidance, particularly in the minds of the client and his advisers. This alone is making for a much more professional approach to enquiry preparation and contract administration.

The range of main options from lump sum to cost reimbursable provides sufficient flexibility for a client to allocate financial risk as he sees fit for his particular circumstances. The list of compensation events makes it clear that each party carries the financial risks associated with his own contribution. The contractor also knows he will be reimbursed if an identified compensation event does occur. This requires the employer to budget for compensation events.

In forecasting the effects of compensation events, the contractor is allowed to include cost and time risk allowances for matters that have a significant chance of occurring, and are at the contractor's risk under the contract. However, alternative quotations for compensation events may be provided by the contractor, on the basis of excluding certain of those risks with the project manager agreeing to pay for them if they do occur. This is one of the ways of managing financial risk available under NEC, which allows the employer only to pay for a risk in a change situation if it should arise.

How NEC deals with quality

Under the ECC the contractor is required to provide the works in accordance with the works information. This document will include a description of the quality assurance system, which the contractor is required to implement. The description can be drafted by the employer's team, or by the contractor in response to a request in the invitation to tender for him to propose a suitable scheme for acceptance by the employer as part of his tender. The works information will also include details of how the supervisor will carry out random checks on the operation of the system, and of the contractor's work.

The PSC requires the consultant to operate a quality management system for his services as stated in the scope. It is likely that he will have proposed the system as part of his negotiations for the appointment.

Quality standards and tests of workmanship and other deliverables will be stated in the works information and scope, respectively. The largest contribution to quality will be in the design itself, standards for which will also be included in the works information and the scope.

Fitness for purpose, or reasonable skill and care

In the ECC, Clause 20.1 states that 'The Contractor Provides the Works in accordance with the Works Information'.

The manner in which the works information is drafted will therefore determine whether this is to be on the basis of fitness for purpose or otherwise. Secondary option M can be included in the contract to reduce the contractor's liability for his design to reasonable skill and care. This would usually apply when he is required to employ a consulting architect or engineer to carry out the design work.

Defect correction

A contractor's liability for correction of defects is linked to his obligation to 'Provide the Works in accordance with the Works Information'. The contractor must do this until the defects date, and in order to obtain the defects certificate, which in turn releases the second half of retention, and the performance bond (if, and to the extent, applicable). The ECC provides a useful clause which allows the project manager to accept defects that are impossible to put right, e.g. a pile that has to be in undisturbed ground and fails its load tests and so cannot be replaced. Similarly, the project manger is given the right to accept the defects that would be disproportionately expensive to correct. In these cases the contractor is asked to propose a reduction in price and/or time to compensate for the defect. This common sense approach removes what under other more rigid forms of contract can be pointless irritations. Furthermore, the contractor does not have to be notified of a defect by the employer's project manager or supervisor. This ensures that the contractor's obligation remains in place, irrespective of what the supervisor might or might not say or do.

How NEC deals with time

Appropriate planning provisions

Each contract in the NEC family contains planning and programme provisions appropriate to the uses for which the contract is designed. The planning procedures in the PSC are different and better suited to the management of such services. In the ECSC, planning procedures are only a requirement of the works information, as the sophisticated procedures in the ECC may not be an essential feature of the management of an ECSC contract.

The programme and planning requirements contained in all the NEC documents are generally more comprehensive than those contained in most other standard forms of contract. This is in response to one of the stated objectives of NEC, namely to introduce modern project management into contractual arrangements.

The NEC planning and change control procedures need to be viewed in terms of a balance between the cost of resources, the potential for savings in the total cost of procurement, and the enhanced control that accrues through their effective application

The programme in the ECC and ECS

Provision is made in the ECC and ECS for a programme either to be identified in the Contract Data, Part Two at award (this means it is submitted with a tender or submitted just prior to contract award), or to be submitted by the contractor after award and within a period stated in the Contract Data, Part One.

The programme is an important document for administering the contract. It enables the project manager and contractor to monitor progress and to assess the time effects of compensation events, including changes to the completion date.

Programme with tender optional

Employers may wish to have programmes submitted with tenders in order to judge whether a tenderer has fully understood his obligations, and whether he is likely to be able to carry out the work within the stated time, using the methods and resources proposed. Any doubts on these matters can then be resolved after submission of tenders and prior to award of contract.

Because of the cost to contractors of providing the required programme with their tenders, employers for their part need to give serious consideration to whether it really is needed from all tenderers. The decision should be based on the criticality of time interfaces and the complexity of resource management, rather than on contract value. Employers also need to recognise that a tenderer's programme will in any case only be as good as the information provided by the employer at tender stage. Generally, a more practical approach is to include in the tender stage the procedures for programme submissions to be called for by the employer, after tender, and then only from tenderers with a chance of being awarded the contract.

Content of the programme

The ECC programming clauses list the information that the contractor is required to show on each programme submitted for acceptance. It consists of:

- dates which are stated in the Contract Data or the Works Information,

- dates decided by the contractor,

- method statements,

- order and timing,

- float and, separately, time risk allowances,

- health and safety requirements, and

- other information required by the Works Information.

In the 'lump sum' priced activity schedule Option A contract, and the target contract with activity schedule (Option C), each priced activity is to be identified on the programme with its start and finish dates shown. In effect, this means that each activity on the programme is priced, and cash flow is demonstrated as a result.

The project manager is required to accept the programme or state specific reasons why it needs to be revised within the time-scale stated in the contract, usually two weeks.

Method statements

Method statements for the contractor's operations consist of descriptions of the construction methods as well as details of the resources, including equipment, he intends to use. Thus any reference in the contract to the programme includes these method statements. This means, for example, that a contractor's quotation submitted in relation to a compensation event that includes a revised programme must also include any revised methods of construction and application of resources. A revised programme is necessary even if only methods and resources are changed without there being any change to the order or timing of the activities.

Time risk allowances

The contractor's time risk allowances are to be shown on his programme as allowances attached to the duration of each activity, or to the duration of parts of the works. These allowances are owned by the contractor as part of his realistic planning to cover his risks. They should be either clearly identified as such in the programme or included in the time periods allocated to specific activities. It follows that they should be retained in the assessment of any delay to planned completion due to the effect of a compensation event, but can be sacrificed by the contractor in an alternative quotation requested by the project manager in terms of Clause 62.1.

The programme is also to show dates by which the contractor requires information, facilities, possession, etc.,

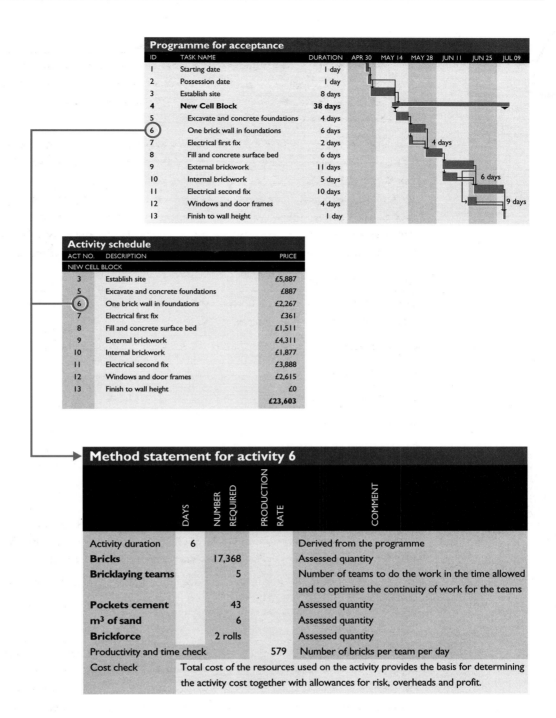

Programme for acceptance

ID	TASK NAME	DURATION
1	Starting date	1 day
2	Possession date	1 day
3	Establish site	8 days
4	**New Cell Block**	**38 days**
5	Excavate and concrete foundations	4 days
6	One brick wall in foundations	6 days
7	Electrical first fix	2 days
8	Fill and concrete surface bed	6 days
9	External brickwork	11 days
10	Internal brickwork	5 days
11	Electrical second fix	10 days
12	Windows and door frames	4 days
13	Finish to wall height	1 day

Activity schedule

ACT NO.	DESCRIPTION	PRICE
	NEW CELL BLOCK	
3	Establish site	£5,887
5	Excavate and concrete foundations	£887
6	One brick wall in foundations	£2,267
7	Electrical first fix	£361
8	Fill and concrete surface bed	£1,511
9	External brickwork	£4,311
10	Internal brickwork	£1,877
11	Electrical second fix	£3,888
12	Windows and door frames	£2,615
13	Finish to wall height	£0
		£23,603

Method statement for activity 6

	DAYS	NUMBER REQUIRED	PRODUCTION RATE	COMMENT
Activity duration	6			Derived from the programme
Bricks		17,368		Assessed quantity
Bricklaying teams		5		Number of teams to do the work in the time allowed and to optimise the continuity of work for the teams
Pockets cement		43		Assessed quantity
m³ of sand		6		Assessed quantity
Brickforce		2 rolls		Assessed quantity
Productivity and time check			579	Number of bricks per team per day
Cost check				Total cost of the resources used on the activity provides the basis for determining the activity cost together with allowances for risk, overheads and profit.

Examples of a programme, activity schedule and method statement, for submission as part of a contractor's tender

which are to be provided to him by the employer. Where there is a large amount of contractor design, it may be appropriate for the works information to ask for more detail to be shown on the programme as the design is developed.

Revising the programme

The first programme, submitted either with the tender or within the time stated after award, is clearly based on the information available to the contractor at that time. The ECC requires that the programme be revised at stated intervals, so that it always constitutes an up-to-date plan for the remaining work. Each revision is to record the actual progress achieved on each operation and the reprogramming of future operations. It should also show the effects of implemented compensation events and early warning matters. If a compensation event affects the timing of future operations, a revised programme indicating the effects is to be submitted as part of the contractor's quotation (Clause 62.2). The revised programme should also show proposals for dealing with delays, defects and any changes the contractor wishes to make.

Benefits of the planning requirements

The contractor places himself at considerable financial risk if he does not provide a realistic programme in the first place and then keep it up to date to always demonstrate his plans for the work not yet done. This is because when a compensation event occurs and there is no up-to-date programme available, the project manager is required to make an assessment of the effects of the event, and will inevitably do so based on incomplete and out-of-date information. It would be very difficult for the project manager to assess time risk allowances for example.

Because the contractor owns the float in his programme in the ECC, there is no motivation for him to provide one programme to the project manager that may 'look good' or suit the project manager's perceived objectives, and then produce another programme for use by his own people. This practice, often used with older more traditional contracts, will be detrimental to the contractor under the ECC.

The project manager, in reviewing a submitted revised programme, would need to note any changes to the dates by which the employer is required to provide information, facilities, possession, etc. He should be prepared to accept a programme with earlier dates if this is acceptable to the employer. However, after acceptance, any subsequent failure by the employer to meet these earlier dates is a compensation event.

The real benefit of the programme and planning requirements of the ECC, apart from the obvious one of always having an up-to-date plan for the remaining work, is in having a firm and factual basis upon which to calculate the cost and time effects of compensation events.

Extensions of time

Under NEC contracts there is no 'entitlement' to extra time in the traditional sense. The contractor (or consultant under the PSC) substantiates the effect of a compensation event on his latest Accepted Programme as it was at the time the event occurred, and forecasts the time implications based on the effect the event has on his critical path network. Time for completion (the Completion Date) can then be extended accordingly in a revised programme. The contract implies that terminal float in the contractor's programme remains with the contractor. If however, the Project Manager is not in a position to extend the time for completion, he is able to ask for alternative quotations from the contractor for finishing on the as yet unaltered Completion Date applicable at the time when the compensation event occurred. Clearly this may entail additional expense for the employer, but at least the employer has that choice.

Acceleration

Acceleration only applies in circumstances where the employer requires an earlier contractual Completion Date, for whatever reason. The Contractor does not have to provide a quotation if he does not wish to, and he is not tied to the compensation event method of assessment. This has nothing to do with making up time for the effects of the Contractor's own delays.

How NEC deals with money

The different payment mechanisms for each of the main options in an NEC document are based on the use of two key terms:

●   the Prices, and

●   the Price for Work Done (or Services Provided) to Date (PWDD).

The first of these two terms relates to the contractor's (or consultant's) prices for components of work, and the second to cumulative amounts due to the contractor (or consultant) for work done by the end of each assessment interval.

NEC documents avoid the confusion, which many traditional contracts contain, between 'cost' and 'expense'. They also do not use terms like Contract Sum and Contract Value. Quite simply, the tendered amount in one of the priced options is the Total of the tendered Prices. When agreement has been reached on this figure, just prior to award, it becomes the Total of the Prices at the Contract Date. As additional work and/or compensation events arise during the contract, the Total of the Prices changes, and therefore always needs to be stated in conjunction with the time at which it is applicable.

Depending on the selection of secondary options, amounts due for retention and price adjustment for inflation are applied to the current Price for Work Done to Date. Delay damages are stated as amounts due per day of delay. Performance bonds, advance payments, etc., are also to be stated as amounts and not percentages. This reduces the uncertainty associated with forecasting total out-turn cost by applying a percentage to an amount that varies. This discipline also requires the employer to think about the amount of retention, delay damages or performance bond he really needs, or whether he needs it at all, rather than just accept or state a percentage, because it is common practice to do so.

There are only two other references to 'money' in the ECC. The first is through the use of the defined term Actual Cost. This definition (explained in more detail below) varies depending on which main pricing option is selected, but relates to defined components of the contractor's costs. The second is the term Fee, which includes everything not covered by the defined components of cost, and is calculated by applying a tendered fee percentage to the total of (the components of) Actual Cost.

Payment mechanisms in the ECC

An assessment for payment is the amount due at the time the assessment is done in terms of the chosen payment strategy. This is done to remove the practice of undertaking only a provisional, usually monthly, assessment with an all-embracing re-measure and 'claim settlement' at the end of the contract.

As an example of how the terminology is applied, in the ECC each term is defined for each main option in Clause 11.2 as set out in the following table:

Option	The Prices	The Price for Work Done to Date
A	AS prices for activities, 11.2(20)	Total of the Prices for completed activities, 11.2(24)
B	BQ rates and prices	Quantities of completed work at BQ rates and proportions of lump sum prices, 11.2(25)
C	AS prices for activities, 11.2(20)	AC paid + Fee, 11.2(23)
D	BQ rates and prices, 11.2(21)	AC paid + Fee, 11.2(23)
E	AC + Fee, 11.2(19)	AC paid + Fee, 11.2(23)
F	AC + Fee, 11.2(19)	AC accepted for payment + Fee, 11.2(22)

AS, activity schedule; BQ, bill of quantities; AC, actual cost.

In the Option B Priced contract with bill of quantities, ECC does not provide for approximate bills of quantities, provisional sums or prime cost sums. This means that Works Information has to be complete, even if it is not final. Consequently, ECC contracts have a clear basis and this reduces problems and disputes.

Time of payment and cash flow

The ECC and ECS certification and payment procedures have been designed very much with a subcontractor's cash flow position in mind.

Assessment dates are set to suit the parties at the start of a contract. These would be set to suit typical monthly measurement or payment dates. Certification of payment by the project manager in the ECC is to be

Emergency projects in South Africa

The South African Public Works Department has been faced with many 'emergency projects' to be undertaken since the first fully democratic government came to office in 1994. These include upgrading 53 police stations, mostly in remote sites, which the politicians demanded should be started immediately. There was not enough time to implement the usual procedure of first appointing an architect to prepare fully detailed designs and drawings for measurement by a quantity surveyor and preparation of bills of quantity adequate for tender purposes. Negotiation was not allowed under Government regulations. Fully open tenders had to be arranged and done so in a manner that would encourage newly emerging enterprises to tender for the work.

Urgency dictated that the scope of work could be determined only after the contracts were awarded and, even then, significant changes occurred as new requirements came to light. The Public Works Department chose to use the ECC, Option E, cost reimbursable contract. This was the only form of contract available that could accommodate all the Government's criteria.

The work was arranged in groups of stations or individual stations, taking account of geographical constraints and the capacity of local contractors. Enquiry documents were prepared by the professional team in a couple of days, and they then provided training for the tenderers in what was required.

Tenders were returned in two weeks. They were evaluated on a system that awarded points for price (the percentage for Working Area overheads and the fee percentage), the contractor's experience, the capability of teams and equipment to start work immediately, and the contractor's commitment to maximising the use of local labour and services.

Work started within a few days of deciding who was to be awarded each contract. The contractor for each project worked with the professional team in determining the starting Works Information. As new information came to light during construction, changes were made, again with the contractor working with the professional team.

During construction, the programme was planned by the Project Manager and the contractor working together using a database of time and cost information that complied with the defined Actual Cost in the contract and so was at the level at which components of construction were either bought or hired. The work activities were detailed on a programme. Each activity was supported by its detailed method statement, which identified the plant and other resources that the contractor planed to use. These descriptions included the quantities of materials required, the number, composition and usage of the labour teams needed, and similar details of the plant and temporary works required. Each resource was then costed at Actual Cost. The aggregate of the required resources for each activity was then summed and smoothed for the whole contract. This provided the forecast planned completion and total cost.

The NEC cost reimbursable contract is not a traditional 'cost-plus' contract. Cost and time control are based on forecasts of the future work to be done, not on historic 'dayworks' records of what was done, which always leads to arguments about whether the resources should or should not have been used. Not all costs are reimbursable. Costs not covered by the definitions in the Schedule of Cost Components are deemed to be included in the contractor's fee percentage. This provision avoids conflicts over what should or should not be reimbursable and makes it abundantly clear that contractors have to accept responsibility for their own construction efficiency.

The contract requires that all amounts included in Actual Cost are at open market or competitively tendered prices with all discounts, rebates and taxes that can be recovered, deducted. This provision limits payments to the contractor to the level determined by the open market.

Working within this contractual framework, the contractors prepared forecasts of the total finished cost in consultation with the Project Manager at predetermined intervals. The contractors also submitted explanations of changes made since the previous forecast. This required and promoted a proactive and forward thinking approach from both the contractors and the professional team.

The provisions of the contract allow for large increases or decreases in the scope of the work. So there is sufficient flexibility to meet the client's requirements, including, if necessary, changing the scope of work so that it fits the budget. All this was achieved without the traditional risk of claims.

All the other provisions of the NEC family of contracts applied. Those that provide for prompt payment to contractors and subcontractors with interest on late or incorrect payments being particularly useful in ensuring that the contractors had confidence in the projects. Also, the cost and time management and the risk allocation between the parties meant that once the scope of work had been defined, the contract could be converted (by agreement) into a lump sum contract or a re-measurement form.

The contracts were completed in about 25% of the normal time under normal Government procedures. Some R100 million (£10 million) of work was successfully completed and benchmarked against open market or competitively tendered prices to ensure that it provided good value for public money. All normal quality standards were achieved, and in most cases bettered. By any standards these projects were unusually successful and the NEC cost reimbursable option was widely seen as being absolutely central to this significant achievement.

Cell blocks at Tsomo Police Station, before and after upgrade

Charge office at Peddie Police Station, before and after upgrade

within one week of each assessment date, and by the contractor, in the ECS, is required within two weeks of each assessment date. Payment of the amount certified is to be made within a fixed period of each assessment date. In the main (ECC) contract, payment is made by default within three weeks of each assessment date unless a longer period (not to exceed five weeks) is stated in the contract data. In the ECS, payment is required within four weeks of the assessment date, unless a different period is stated in the subcontract data.

There is no link between the two contracts in the form of a 'pay when paid clause'. The certification and default payment periods in the ECS are one week longer to avoid the main contractor having to finance payments to subcontractors for main contractors on Option A, B and F contracts. Main contractors on Option C, D and E contracts are required to pay subcontractors first before including these amounts in their own assessment of the amount due from the employer.

Cost to employers of late payment

It is often the case that the major value portion of contract works is carried out by subcontractors. If either the employer or the main contractor increases the above default payment periods, it will not be difficult for the employer to estimate the net cost to him of financing the effect of the delayed payments that a contractor, and particularly a subcontractor, will need to allow for in their prices.

Assuming that many employers can finance their projects at a rate 4% below that at which most contractors and subcontractors borrow short-term operating capital, a consistent habit of early and timely payment by employers could attract a substantial reduction in tendered prices.

In all NEC documents, interest is payable on amounts that should have been certified and on amounts that are paid later than within the time stated. Clearly, it is in the employer's interest that all subcontracts are based on the ECS, that both the employer and the contractor resist the temptation to extend the default payment period, and that payment is actually made within the stated times.

Cost reimbursement option

A cost reimbursable contract is ideal when the definition of the work to be done is inadequate even as a basis for a target price, and yet an early start to construction is required. In such circumstances a contractor cannot be expected to take cost risks other than those that entail control of his employees and other resources. He carries minimum risk and is paid Actual Cost plus his tendered fee, subject only to a small number of constraints designed to motivate efficient working.

In the ECC Option E contract, Actual Cost is defined as the amount of payments due to subcontractors for work that is subcontracted and the cost of the components in the Schedule of Cost Components for work that is not subcontracted, less any Disallowed Cost which is also defined.

Clause 52.1 further states that all contractor's costs which are not included in the defined Actual Cost are deemed to be included in the *fee percentage*. Amounts included in Actual Cost are at open market or competitively tendered prices, with all discounts, rebates and taxes that can be recovered deducted.

Hence, a number of specific controls are available to the employer in the payments he makes under the ECC 'cost reimbursable' Option E, namely:

- Cost is defined in terms of payments due to subcontractors and for work not subcontracted. It is defined as a set of listed components (and no other) in the Schedule of Cost Components. In both cases, amounts admitted for payment (by the project manager) in terms of those definitions have to be at open market and competitively tendered rates. Anything not covered by the listed Schedule of Cost Components, or payments to Subcontractors, is deemed included in the *fee percentage*. The *fee percentage* is tendered.

- The Fee is the fee percentage applied to the subtotal of Actual Cost, and the total amount due is Actual Cost plus the Fee.

There are difficulties in agreeing assessments on the basis of open market competitively tendered rates, especially in remote areas with limited suppliers of materials and resources. However, even allowing for these difficulties, the degree of control still provides employers with a firmer sense of security about final out-turn cost than they are used to in traditional cost reimbursable contracts. The contractor has an incentive to monitor his productivity to ensure he is within local market-related tendered prices.

The benefit of the system when an early start is required is that the only details required from each tenderer are his overhead percentages, some off-site design and manufacture and fabrication rates, and his *fee percentage*. More emphasis would be given to the nature and experience of tenderer's resources available to do the work, and schedules requesting such detail would be included by the employer in his invitation to tender. However, almost no details about the work to be done are necessary at tender stage, other than specifications of constraints within which the contractor is to carry out the work and, of course, workmanship specifications. Tenders can be returned in a matter of a few days and assessed with equal speed. The contractor can start work on site as soon as the first details become available, however small the scope of work.

It is expected that this form of contract will have considerable appeal in the reparation of war damage and damage arising from natural disasters.

How NEC deals with changes

Dealing with change has always been the Achilles' heel of construction contracts. The system of change management in NEC overcomes the weaknesses all too evident in traditional contracts, in the fairest possible way to both parties. The procedure for dealing with change is the same, irrespective of pricing strategy and the nature of the event giving rise to the change. However, it is applied differently depending on which main option is used. It is based on the following principles.

Identifying change

A list of 'compensation events' is provided. Compensation events relate to:

- things which the employer, project manager and supervisor do to the contractor during the life of the contract;

- stated risks of loss and damage carried by the employer, the effects of his design and the use of the works;

- risks arising from adverse physical and unusual weather conditions which are shared between the parties; and

- other events (such as changes in the applicable law) depending on the selection of option clauses.

These are the only events for which compensation can be claimed and they include the employer's breach of contract. Hence, the contractor's risks are clearly identified to the extent that those events which are not listed in the contract are to be allowed for within his prices (if Options A to D apply). A contractor is therefore relatively indifferent to the changes that an employer might make, and can tender accordingly.

Compensation for changes

The list of compensation events (18 in the ECC) includes matters such as changes to the works information and late supply of items that the employer is to supply. The occurrence of such events leads to the contractor being compensated by the employer for his forecast additional cost arising from the event, and time to the extent that planned completion is affected by the event.

Hence, all events for which the contractor may be entitled to extra payment are listed in one place, and the procedure for assessing the time and cost implications is identical for all events.

Reductions to the scope of work would be assessed and priced in the same way as additions; however, the compensation event procedure does not permit the time for completion to be shortened as a result of the change to scope of work.

The pricing of compensation events on the basis of carefully defined Actual Costs plus a predetermined fee, is designed to be fair to clients and contractors. Bill rates can be used if both parties agree (when Option B applies), but the use of actual cost means that major changes will not unfairly reduce or increase a contractor's profit margins.

ECC allows for contractors to provide alternative quotations for compensation events. A common example is to have alternative prices, one allowing for an extension of time and the other for no extension of time. The project manager, in consultation with the employer, can then select the best-value alternative. This provision often leads into innovative ways of dealing with a change, because the contractor and project manager can consider various ways of dealing with any given problem.

NEC tries as far as possible to remove the need for subjective judgements. For example, in the ECC adverse weather conditions are dealt with in terms of hard numbers in place of the usual vague criteria. This makes it more likely that claims are based on factual information rather than opinion. Again, this helps avoid silly arguments that demotivate everyone involved.

Valuation of the change

Valuation of any claim for compensation is based on a 'cost plus' principle. For work not yet done it is valued as a forecast by the contractor of the cost of the effect of the event plus the fee. For work already done it is valued at proven actual cost plus the fee.

Cost, depending which main option is being used, is determined by a set of components relating to the cost of people working within the working areas, equipment which the contractor uses to provide the works, plant and materials built into the works, charges for site services and certain overheads, off-site manufacture and fabrication and design charges, and their associated overheads. Only the off-site rates and their overhead percentages are provided by the contractor at tender stage. The other components of cost are substantiated by the contractor at the time of submitting a quotation for the effect of the change. In main Options C, D, E and F, components of actual cost also include payments to Subcontractors. Actual Cost is further required to be competitively tendered, or at open market rates.

The fee to be added (cost plus fee) is the total of the above components of cost multiplied by the contractor's tendered fee percentage. The contract deems that all the contractor's costs that are not included in the defined list of components of cost are included in the fee percentage. The tendered fee percentage would most likely include profit and the proportion of (off-site) corporate overheads charged to the contract.

Hence, if the project manager accepts the contractor's notice of the occurrence of one of the identified compensation events, or requires a compensation event (e.g. a change to the scope of work) to be valued before implementation, the contractor is required to value it as:

● the Actual Cost of work already done,

● the forecast Actual Cost of the work not yet done, and

● the resulting Fee.

Revising the prices

Having agreed the effect of the change, the prices are changed if Options A or B apply. The target would be changed if Options C or D apply, and only the total budget for providing the works would be changed if Options E or F applied. In the case of Options C, D, E and F, all payments to the contractor are made at Actual Cost plus Fee (being essentially cost plus contracts), which is why compensation events only affect the target or budget.

Time effect of a change

The time effect of a compensation event is taken account of when the event arises, since each quotation submitted by the contractor is required to include a revised programme showing the effect of the event on the remaining work. Delays are calculated by the contractor as the effect of the event on the contractor's planned completion, and added to the Completion Date. Terminal float remains with the contractor, and other time risk allowances, if identified on the contractor's latest Accepted Programme, are also retained unless sacrificed in an alternative quotation. The project manager may ask for alternative quotations for different ways of dealing with the compensation event, including asking the contractor to value the cost to the employer of completing the works in the same time, should it not be possible for him to extend the time for completion. This should not be confused with acceleration in terms of Clause 36, which is used to bring the Completion Date forward and is not part of the compensation event procedures.

Acceptance of a quotation

The project manager, on receipt of the contractor's quotation, checks that only the listed components of defined Actual Cost have been included, that the extent of them is justified or proven (in the case of work already done), and that the rates used are open market or competitively tendered.

On acceptance by the project manager of the contractor's quotation, the prices are changed. In the case of work not yet done, the contractor is then at risk (in Options A and B) to complete the work required by the compensation event for the agreed price.

Timing and administration

There are strict time limits at all stages of the procedure designed to encourage the valuation and re-programming made necessary by a change to be undertaken at the time it occurs, when records are available. This is important in preventing unresolved matters being allowed to accumulate and gathered into a massive claim at the end of the contract, as is common under traditional practice.

There are no monetary limits beyond which the process of change is reconsidered. This is because the contractor's quotations, which the employer is free to accept or reject, will reflect the consequences of substantial changes if they adversely affect the contractor's ability to execute them within a reasonable price and time.

No provisional sum or day-work items

The first compensation event, 60.1(1), states 'The *Project Manager* gives an instruction changing the Works Information'. This includes 'day-works' that are valued as stated above, either at proven Actual Cost or as a forecast Actual Cost for work not yet done. This gives a greater degree of cost control for the employer. Hence, a traditional day-work schedule is not required in an ECC contract.

Where there is uncertainty about the scope or inclusion of certain items of work, it is traditional practice to include a 'provisional sum' for these in contracts using a bill of quantities. In an ECC contract (Option B), the work would be described and billed as best as possible based on the information available at tender stage. When the final scope of work is known, if it turns out that the items in the bill of quantities do not fairly describe it, the project manager 'instructs the change' as a compensation event and the time and cost effects are determined as described above. The key to development of the works information in all NEC works contracts is to ensure that it is complete even if it is not correct at the time of contract award. The effect of the change is then determined at Actual Cost plus the Fee.

Some words of caution for contractors and employers

It can be seen that the change management procedure in NEC documents is very different from traditional practice. Training in its application is essential. Initially, contractors will be exposed to the employer's project manager's knowledge of market-related cost, time risk allowances and programmes, and the ability he has, or has been given by the employer, to deal with compensation events within the prescribed times.

The employer's risks associated with change in an ECC/ESC-based contract are likely to be better controlled

A typical installation built by Sasol, the South African petrochemical giant. Today Sasol uses the NEC for the greater part of its new work.

Majuba Power Station, the last in a series of eight similar six-unit generating stations completed by Eskom since 1980. In 1992 Eskom standardized on the use of the NEC for all new work

than with traditional contracts, as they are clearly highlighted for all to see. From experience to date, financial risks are dependent to a very large extent on both the quality and completeness of the works information. Spending extra time and effort on the works information before going out to tender will pay great dividends. This is true with any form of contract, but the ECC documents appear to be forcing a long overdue need for such improvements, from which the whole industry must benefit.

How NEC deals with disputes

The early warning mind set

One of the fundamentals of good project management is that members of the project team collaborate with each other regarding anything that might have an effect on the time, cost or performance of the project in order to shrink the risks inherent in any engineering and construction activity. This requirement is built into NEC contracts, and there are potentially severe consequences for the contractor if the warning is not given, and for the employer if the project manager does not deal with them at the time they are given.

If the contractor does not provide the required early warning, any subsequent compensation event that he notifies will be assessed for both cost and time implications as though the early warning had been given. The contractor therefore risks not being compensated for substantial work if the warning is not given. If the project manager does not provide or deal with early warnings, as he should, he loses the flexibility of alternative ways of dealing with the matter, as well as an opportunity to provide the employer with choices before work is done. The resulting compensation event could also be much more expensive for the employer than might otherwise have been the case.

Not giving early warnings, or not dealing with them when they are given, is the first seed for a potential dispute. This is because the resulting compensation event assessment now has an element of bad taste, or 'win-lose' instead of 'win-win', which is neither good project management nor part of the partnering ethos. The simple way to avoid a dispute is not to plant the seed for one in the first place – give an early warning.

The project manager needs to act as stated in the contract

From experience to date, the other main contributor to disputes is the project manager not acting as he is required to do in the contract, or in the time required. Employers should act swiftly in this case and replace the project manager with somebody who can manage proactively before relationships become strained. The problem could also be the employer himself not providing the project manager with the decisions he needs in the required time.

Adjudication and the *Adjudicator*

Should disputes arise, the NEC documents require that the matter be referred to an independent adjudicator who will give his decision on the dispute with full details of the time and money consequences. The adjudicator is appointed jointly and paid in equal shares by the parties, but only paid when required to apply time and resources to notified disputes. He is not paid a retainer, and does not become involved in the project until a dispute arises. The adjudicator is named at tender stage and appointed at the same time as the contract between the parties to the main contract is concluded.

In complex contracts, the adjudicator could be a panel of experts, depending on how the parties agree to use the process at the time of contract award.

The parties are required to make their submissions to the adjudicator within fixed time periods, and the adjudicator is required to make his decision, which is binding, within a fixed time of receiving the submissions. All the time periods are shorter than traditionally has been the case. If either party is dissatisfied by the adjudicator's decision, the dispute may be referred to either arbitration or court as pre-selected in the Contract Data.

The NEC documents take account of both UK and international application by having two sets of clauses dealing with the adjudication process. This is because, in the UK, adjudication has to comply with the requirements of the Housing Grants, Construction and Regeneration Act of 1996.

Selection of the adjudicator is critical, and users of the NEC are advised to use only adjudicators listed by recognised accreditation bodies, such as the Institution of Civil Engineers. Whilst it is intended that the adjudicator is not a lawyer, he will certainly need to be sufficiently experienced in matters of law at least to know when to seek such advice on behalf of the parties. It goes without saying that he will also need to have a sound knowledge of most of the NEC documents and how they are intended to operate.

Implementing NEC throughout an organisation

Implementing NEC throughout an organisation is always more difficult than on a project-by-project basis. The following issues have been found to arise in most organisations, whether the organisation is in the public or private sector.

The decision to implement NEC must be taken at the highest level of management to obtain the necessary buy-in and commitment.

The decision to implement NEC should be taken at management board level. A phasing in period of a year should be allowed for the changeover from the old to new conditions of contract, after which use of NEC should became compulsory. This is much better than an organisation claiming to implement NEC, but on a voluntary basis. If done on a voluntary basis, the change takes much longer, chaos reigns as staff battle with two different approaches to capital expenditure, and resistance hardens amongst those who have a vested interest in the system being replaced.

It is necessary to check whether organisational (or State) procedures are compatible with the requirements of single point accountability, a project management approach and the NEC.

Corporate directives or state tender board rules often need to be changed to allow the delegation of decision-making to the project manager. Large organisations tend to be structured with separate commercial and technical divisions, with the authority to make cost and time decisions vested only in the commercial division. This simply will not work with the NEC requirement for rapid decision-making by the employer. Project managers need to be allocated a budget for their project, which includes justified allowances for compensation events. They also need to be delegated the necessary authority to be able to act effectively as the agent of the employer in terms of the contract.

Additional documentation should be prepared to match the NEC in style and terminology.

The published NEC documents are not supplied with pro forma conditions of tender, bonds, parent company guarantee, and certificates for the administration of the contract. While these are probably best drafted on an organisation-specific basis, they do need to integrate with the NEC documents in style and terminology, and be drafted without ambiguity. All such pro formas need to be available on an organisation-wide basis at the time of implementation.

Tenderers will probably qualify their bids in respect to overall limitation of contractor's liability, force majeure, damage to employer's surrounding property and latent defects.

Tenderers, especially large, multi-national contracting and consulting organisations, require specific limitation of their liabilities to be included in the contract. Whilst such limitations are to some extent included in the NEC documents, they are not sufficient. Limits to overall liability to the employer, damage to employer's surrounding property and latent defects will need to be addressed by the inclusion of the additional conditions in the secondary Option Z.

Only experienced draftspersons should be used for the complete and careful preparation of Works Information, Site Information (ECC) and the Scope (PSC).

The development of other contract documentation, particularly works information, is a job for experts. Such expertise is not usually found in large client organisations whose core business is not engineering and construction. Experience has shown that as much as possible of the organisation-specific constraints and procedures which it is necessary to include in any contract should be standardised to ensure correct integration with the particular NEC contract being used. Technical staff need to be trained in how to draft specifications, as well as how to express the employer's requirements fully and unambiguously. Cross-checking of technical and commercial specifications for potential ambiguity and inconsistency must be undertaken before documents are issued for tender, and again before contract award.

Recognise the importance of training and stakeholder management.

Liaison, in the form of training and technical assistance regarding the NEC with prospective contractors, subcontractors and suppliers tends to be good in parts of a large organisation and non-existent in others. The latter case usually makes for a badly run contract, so training should always be provided. There is a tendency to provide training to contractors and suppliers for the first few projects, after which subsequent newcomers or first-time NEC users are left to their own devices. The resultant confusion and disillusionment (of the contractor) can affect the client's project to a far greater extent than the cost and time involved in adequate training.

It may be necessary to form a panel of adjudicator's.

Organisations letting large numbers of contracts may find it impractical to appoint an adjudicator at the start of each NEC contract. Some organisations have resolved to make reference in the Contract Data to a panel of pre-selected and suitable adjudicators. Should a dispute arise, the contractor (or consultant) is then given the choice of selecting any one member of that panel who is available to act in the required time. A contract with the selected adjudicator is then prepared using the NEC Adjudicator's Contract.

Chapter Three

CASE STUDIES OF NEC IN USE

● **ABSA Bank Towers in Johannesburg**

● **GDG Management**

ABSA Bank Towers in Johannesburg

- NEC helped a large, innovative building project undertaken in a very difficult economic and political environment to achieve its time and cost targets.

- The interests of all the internal and external stakeholders were explicitly taken into account in all major decisions.

- The project was led by experienced project managers who used NEC because it supports best practice project management.

- The NEC programme provisions provided a robust basis for managing the project.

- The best firms, rather than the lowest bidder, were carefully selected for each contract and each subcontract.

- Workshops and training in NEC were provided for all the many firms involved in the project.

- Teamworking quickly became a key feature of the project team's approach.

- Value engineering was used to optimise the design within carefully defined cost constraints.

- A period was allowed between appointing the contractor and start on site in order to review and improve the constructability of the design.

- The almost continuous design development, which is characteristic of large individually designed building projects, resulted in many changes that required the early warning and compensation event procedures to be streamlined.

- The quality control systems developed and improved in response to lessons identified at three-monthly reviews.

This project, planned and completed in 39 months, provides a US $65 million new bank headquarters which it is hoped will encourage the revitalisation of Johannesburg's central business district. The new building covers two city blocks, has a total floor area of 85,460 m², arranged in three basement levels of car parking, two ground floor levels and six upper floors built around a central atrium. It provides modern accommodation for over 2000 ABSA Bank staff who were previously spread among five other buildings.

NEC was used on this important building project, which involved a considerable amount of change and design development in a difficult environment, and yet was finished on time and within budget. This achievement is unique in the recent experience of the South African building industry. The project demonstrates how the NEC approach, faced with the high number of compensation events typical of large, high-prestige building projects, encourages the project team to work co-operatively and adopt many of the features found in well-developed partnering.

The client set stringent requirements for the project and established a system of measurements to determine how effectively they were being met. A firm of professional project managers, Barrow Projects (Pty) Ltd, was recruited to take primary responsibility for delivering the project. The client, ABSA Bank, has no hesitation in confirming on the basis of objective measurement that project management exceeded their expectations.

Barrow Projects recommended that NEC be used because it supported their own well-established project management system and encouraged co-operation. At the time, there was little experience of NEC in the South

African building industry, so the project manager warned that a major culture shift from traditional practice would be involved. ABSA accepted the project manger's recommendation, and NEC training was provided for all the firms involved.

The design was produced by the client's professional team on the basis of NEC Professional Services Contracts. The client and project manager decided to appoint four *Supervisors* under the main ECC construction contract instead of the usual one, one supervisor from each of the four main design consultants. Each supervisor was responsible for the design continuity and quality of that part of the works designed by his firm, for which they carry professional design liability. This arrangement worked well and did not gave rise to confusion over responsibilities or gaps in the quality control system, which was initially thought to be a concern.

Two workshops were held for the joint client (the employer) and professional project team at a very early stage. All the firms involved were represented at the workshop by their highest level of management, ensuring that they understood and would support the use of NEC. Under the positive leadership of the client, all members of the professional team, sceptical at first, placed themselves and their firms fully behind the intent of the NEC system from the outset. Whilst the industry focused on the 'new' management system for this project, there were also numerous technical advances made in the field of building energy-management systems under the leadership of architect, TC Design Group.

With the workshops creating a new culture of teamworking and innovation from the outset, there were innovative value-added spin-offs into many other aspects of the project and operations of its participating firms, which would probably have never been considered under previous ways of working.

Contractors who pre-qualified for invitation to tender were first provided with the employer's business case for the project and then required to attend, again at their highest level of management, an ECC training course. This took place just before tender documents were issued and was also attended by members of the professional team to provide continuity and start the team-building process.

Construction work was provided through nine contracts, by far the largest being for the building. The other prime contracts were for furniture, fixtures, operating equipment and artworks. The tender documents for the building resulted from a disciplined design process led by the project manager. Value engineering was used to optimise the design within cost constraints. The resulting design, cost plan and programme became the project manager's baseline document in controlling subsequent changes. As a result, the tender documents included drawings and specifications providing a virtually complete description of the building with the matching bill of quantities.

The main building contract provided a construction period of 27 months. It used ECC Option B, Priced contract with bill of quantities, with the important provision that the work was to be completely re-measured to determine payments and the final account. This is an aspect of normal South African practice that is not part of the ECC approach and generally should not be necessary.

The successful contractor, a joint venture between Murray & Roberts and LTA, was given a month before moving on site to go through the design with the professional team. The contractor provided many suggestions for making the design easier to construct and final design details were prepared on the revised basis. One major example is that the original design used normal brickwork, but the contractor proposed replacing this with brick-faced concrete cladding panels. This was accepted by the professional team, which defined the design criteria and worked with the contractor on the re-design, which produced the same costs, ensuring more reliable quality, greater certainty of meeting the programme and a better looking building.

NEC procedures for maintaining up-to-date accepted programmes for all the required work in both ECC and PSC contracts gave the project manager a robust basis for integrating all the work. For example, he used them to maintain up-to-date schedules, which ensured that designers knew exactly what they had to produce to meet the needs of the construction process.

There were 56 subcontractors, appointed in essentially the same way as the main contractor. This included training courses for all the firms invited to tender. The courses were attended by representatives of the

ABSA Bank Towers in Central Johannesburg, a $65million prestige building planned and completed in just 39 months under the NEC

main contractor and professional team to provide continuity and build on the culture change already underway. The training took place before the invitations to tender were sent out, and had a clear and positive effect on the quality of the tender information provided by the subcontractors.

The project team put a lot of effort into selecting the best subcontract tender, which was often not the lowest. This was done on the basis of tender reports produced jointly by the project manager, architect, quantity surveyor and main contractor. These were reviewed at meetings, which also included the subcontractors, as necessary. When a subcontract tender was accepted, there existed a good understanding of what had been agreed and the idea of co-operative working was already in place. This gave several benefits, including faster starts on site, fewer queries and fewer problems. Subcontractors were generally prepared to accept small changes without turning them into compensation events. The fact that the bill of quantities was being re-measured dealt with many of these, but there was also evidence of a willingness to join in the team effort to produce a good project on time and within budget. Substantial changes could be

ABSA Bank Towers provides high-quality accomodation for over 2000 staff

accommodated, agreed early and paid for as the work was underway, so the subcontractors' cash flow was improved and final accounts were settled faster than with traditional approaches.

The project manager held weekly meetings with the main contractor and with the professional team to deal with all outstanding early warning and compensation events. These meetings included the employer's representative, who was responsible for ensuring that decisions required from the client were made in due time. Similarly, the contractor held weekly meetings with subcontractors affected by compensation events currently under discussion. This regular pattern of meetings ensured that, although there were close to 1000 compensation events, they were agreed approximately within the times allowed in the contract. This is widely seen as having provided an unusual degree of certainty over the outcome of the project, and was achieved without recourse or threat of recourse to the adjudicator.

The compensation events generated a mass of paperwork for the contractor, but he gradually recognised that to a large extent this was under his control. In most cases it is the contractor who decides that something should be treated as a compensation event. He realised fairly quickly that it is sensible to absorb low-value items because the paperwork costs more than the likely extra income. This realisation is all part of the contractor developing a more professional approach, which accepts that he is the expert in construction matters and should be able to stick to a budget. Partnering engenders this same approach in the whole project team, concentrating their efforts on meeting the employer's requirements rather than creating paperwork.

The procedures had to deal with some substantial changes to the employer's requirements resulting from the late acquisition of land and the need for a new Dealers' Room incorporating state-of-the-art technology. The risks involved in making these changes were carefully evaluated under the direction of the project manager. The NEC procedures allowed him to seek alternative quotations from the contractor for different ways of dealing with a compensation event, which helped ensure that completion was not delayed.

With the well-disciplined controls afforded by the compensation event procedure firmly in place, the employer was aware of the likely final cost of his project at all stages. As compensation events for additional cost were processed, the employer, in co-operation with the professional team and main contractor, sought other compensation events that provided savings. In this way the client's original budget was never changed, and the final cost was 4% less than budget. Apart from the initial extension of time caused by late access to the site, the main contractor's completion date was never adjusted in spite of the nearly 1000 compensation events. The contractor's quotations for compensation events were always based on maintaining the original completion date.

Quality management began by carefully defining the employer's brief. This provided the key document in quality plans which were drawn up for every firm involved in the project and included in contracts. Performance against these quality plans was formally reviewed every three months to identify problems and look for ways of solving them. On occasions this led to the employer and project manager agreeing that they needed to behave differently because they were inhibiting the quality delivered. In other cases consultants, contractors or specialists needed to work differently, and occasionally training in quality management had to be provided.

An important feature of the project was an early recognition that in such a significant project there are many stakeholders with the power to disrupt or help the project. These included local politicians, local authorities, utility providers, neighbouring owners, trade associations and ABSA Bank's customers and staff. The Bank's Senior General Manager responsible for properties and logistics was given responsibility for managing all the stakeholder interests. He did this by agreeing and maintaining a common shared vision of the project, which took account of all the issues raised by stakeholders. This was actively explained and promoted to all project team members, so they knew how they should work and behave. In this way many sensitive stakeholder issues were identified and dealt with in a proactive, non-confrontational manner.

At the end of the project there was wide agreement between all the parties that NEC made a significant contribution to its success. This is most generally attributed to the NEC approach of requiring everyone to make decisions early and co-operate in resolving problems, so people came on site knowing what work they were supposed to be doing to an extent that is very unusual. They will all welcome NEC being used on future projects.

Whilst the project managers walked off with Project of the Year 2000 awards from the Project Management Institute of South Africa, and ABSA Bank received substantial recognition for its leadership and innovation, the case study revealed that none of this would have been possible without the full support they received from all members of the professional team, the main contractor and all subcontractors. It was a perfect demonstration of the benefits of a partnering ethos.

Picture acknowledgments

Courtesy of Andrew Baird

GDG Management

- NEC helps control difficult, individually designed building projects, which traditionally often suffer from time and cost over-runs.

- Effective project management identifies the stakeholdes and their critical success factors and then delivers the required value at the lowest cost.

- Detailed risk management is central to achieving controlled projects.

- The essential roles and responsibilities are carefully defined and then the best firm available is selected for each one.

- NEC priced activity schedules provide a robust basis for best practice project management.

- Project team members are initially employed as consultants until there is sufficient information to form robust construction contracts.

- Works information in construction contracts must be complete even if it is not final.

- It takes time and several projects to learn how to make the best use of NEC.

- NEC leads naturally into partnering attitudes.

- Firms new to NEC need training, coaching and guiding.

GDG Management are project and construction managers based in the UK who to date have used NEC on 10 projects, which has resulted in them setting up over 100 contracts between their customers and consultants or contractors. GDG have a well-developed and disciplined approach to project management in which NEC is an important tool because it is recognised as providing contractual support and gives legitimacy to their tough approach.

GDG are employed by their customers under an NEC Professional Services Contract using Option A, Priced Contract with Activity Schedule. The scope of services they provide is project management, construction management, or design and construction management, depending on the nature of the project and the level of control and risk the customer feels happy with.

GDG's work principally consists of complex and demanding one-off building projects. These are projects which under traditional approaches often result in cost and time over-runs. GDG have developed an approach that helps them give customers certainty of outcome. It is applied throughout projects and comprises the series of specific actions (see box).

Identify critical success factors by which the outcomes of the project will be judged

Look at risks involved in achieving the success factors

Consider who is most suitably qualified and best able to manage each risk

Identify the co-ordinated set of roles and responsibilities needed for the project team to handle all the risks

Ensure that the customer pays for no more or less than is required to deliver the facility

Main stages in GDG's project management system

GDG's structured approach to project management

Critical success factors

The first stage in the approach is to carefully define what the customer wants and needs. Once it is agreed that a construction project is needed, the first action is to identify the stakeholders. Having done this, GDG next discuss the project with each stakeholder to establish key needs and wants. These are carefully defined as key success factors and agreed by getting each stakeholder to sign a formal statement of those which apply to them.

Key success factors are the criteria used to judge the success of the project. GDG use brainstorming to help identify the key success factors. In doing this they often suggest tentative factors to provoke debate and so establish what really is critical. They find it helps to discuss the customer's brand image and get him to describe what view they want their customers to take. The key question is: What provides value for the customer?

For example, in the case of a university setting up a new business school to produce Harvard-quality graduates, GDG determine exactly what this means in terms of the type, style and quality of accommodation. For a health club needing a crisp image in the reception area and other public spaces, they determine exactly how this will be evaluated by the customer and every other stakeholder who will make judgements about the new facility. Food production facilities need a substantial, clean floor which is inspected in quite specific ways. For some customers, completion by a specific date is crucial, while others have firm budgets. GDG establishes these critical success factors right at the start of projects and gets them signed off. Some take the form of constraints, e.g. completion by a specific date.

GDG project managed the redevelopment of HMV's landmark Oxford Street store, working to ensure the architect's vision was fully realised

Once the critical success factors have been defined, there is a clear basis for making decisions about the project. Essentially GDG's approach defines success in terms of the ratio of value to costs. Given this framework, success comes from defining carefully what represents value for the customer and then delivering the defined value at the lowest cost whilst staying within the defined constraints.

Risk management

The second main stage in the approach is to identify and define in detail the risks involved in achieving each of the key success factors. Common risks include ground conditions, the state of existing services, foundation design, the co-ordination of complex services, weather, the overall cost and time.

GDG then decides who is best able to manage each risk. The clarity of the NEC Engineering and Construction Short Contract's approach to weather helps in dealing with risks. This requires contractors to allow for being prevented by weather from carrying out work on up to one-seventh of the total number of days between the starting date and the completion date. So up to one-seventh is at the contractor's risk and any days above one-

Interior and exterior views of the completed HMV Oxford Street store

seventh are at the customer's risk. GDG tries to define the management of all risks with this same clarity.

The outcome of this stage is a schedule defining all the risks, stating the extent to which each is to be managed by the customer, consultants or contractors. The schedule of risks is included in all the contracts used to establish a project team. The schedule of risks is crucial to the approach because experience has taught that trust and co-operation begin with clarity.

Selecting project teams

The next stage in the approach is to identify the roles and responsibilities needed to meet the key success factors and manage the risks. It may be decided that one firm of architects should design the overall concept, taking account of the customer's brand image and current trends and fashion. But a different firm may be needed for the production drawings and specifications because this requires different skills.

In this way firms competent to undertake each distinct role and responsibility are selected. For most projects, firms initially are employed as consultants because there is insufficient information to produce clear and complete works information. On the type of projects they deal with and for the kind of customers they work with, GDG consider that the NEC's target and cost reimbursable options create too many risks. They like the certainty provided by a priced activity schedule linking the required work to cost and time in detail.

It follows that the first step in setting up project teams is to select consultants. Firms are invited to tender on the basis of full and clear descriptions of the scope of services. Once selected, the consultants work together using brainstorming workshops to find the best ways of meeting the customer's critical success factors.

The initial stage may require firms with construction expertise, and they are employed as consultants using the NEC Professional Services Contract. For example, on a project in which the design of the roof was critical, a roofing specialist contractor was employed as a consultant to develop the design. The resulting design then formed part of the works information used in appointing a contractor on the basis of the NEC Engineering and Construction Contract.

Once sufficient information is available to produce complete works information, competitive tenders are invited from selected contractors to undertake each of the roles and responsibilities. This might include contractors involved in the design stages, because their consultancy work is complete and so there is no conflict of interests.

The most important part of the approach at this stage is to ensure that the works information is complete. A key principle in the approach to works information is to ensure that it includes all the design decisions that consultants are going to make, and that any remaining design is made the contractor's responsibility. The extent of consultants' design varies from project to project, depending on the risks. The consistent principle is that the works information is complete in the sense that the consultants do not need to add further information later. Changes may be necessary, but GDG know it is vital if they are to maintain control of projects that changes are related to complete works information. This tough principle stems from experience of traditional approaches, which shows that if contractors are asked to tender on inadequate works information they bid low to get the work and then concentrate on manufacturing the basis for claims. Those who stay in business are very creative at this.

A further stage in setting up the project team comes when the selected contractor appoints subcontractors. GDG encourages contractors to use the same careful approach in selecting subcontractors as that used in selecting the contractor.

These distinct stages in building up the project team mirrors the progress of all projects as they move from broad plans and high risks to detailed plans and low risks. On very complex building projects, a construction management approach is used initially until sufficient decisions have been made to allow a priced contract to be formed on the basis of defined risks.

This systematic approach of setting up the project team in step with defining the risks takes time. However, simultaneous engineering approaches that bring the whole team together in a project office right at the start of a project and begin work on site just a few days later are seen as too risky. The approach aims to deliver value and certainty in a carefully managed approach, rather than trying to begin work on site at the earliest possible date.

Priced activity schedules

The approach to project management is based on priced activity schedules. Consultants and contractors are required to submit priced activity schedules along with their tenders, in sufficient detail to make clear how it is proposed to manage each of the risks they have been allocated. All the deliverables are listed, e.g. design consultants list all the drawings and specifications they will produce.

Experience of using other forms of contract, principally JCT,[1] suggests that at tender stage consultants and contractors often have a poor understanding of the risks involved in their work. Priced activity schedules are used because these force consultants and contractors to think carefully about all aspects of their work before they submit a tender. This often requires contractors to seek detailed proposals from subcontractors during the tendering period, so that the preferred proposals can be incorporated in the priced activity schedule that forms part of their tender. The priced activity schedule is reviewed and agreed before appointing any consultant or contractor, because this establishes a clear, transparent basis for working in a spirit of mutual trust and co-operation.

The priced contract with activity schedule option is used for all projects because it provides the tools needed to give customers certainty of outcome. Other options might in theory be more appropriate for some projects, but there are great advantages in working with one well-understood approach that above all makes the risks clear and transparent.

As an example of the benefits of insisting on detailed information from consultants and contractors before they are appointed, a design consultant on one project included dealing with compensation events as an activity in the activity schedule submitted with its tender. This led to a discussion with the customer about the degree of change they envisaged, which produced a saving on the consultant's fees equal to over 50% of the consultant's construction stage fee. A further benefit of the discussions was that it identified misunderstandings about which risks were carried by the customer and which by the consultant. These were sorted out so the project began with a clear understanding of who carried each risk.

Experience shows that priced activity schedules help identify the effects of changes to the scope of a consultant's services or a contractor's works information, and so minimise the risk of cost or time surprises.

---

1 Authors' note: JCT refers to the Joint Contracts Tribunal series of standard forms of contract.

As a result, in the great majority of cases the NEC procedures by which compensation events are evaluated and agreed confirm the project manager's judgements about the right costs and times. This encourages proactive project management.

For all these reasons the key management tool is the priced activity schedule. NEC supports this key management tool with a well thought out contractual framework.

Adopting NEC

On the first project where GDG used NEC most of their effort went into learning the new terminology. Once work began, NEC was used in name only as everyone reverted to traditional behaviour. Compensation events went ahead without prior evaluation and were priced after the event using essentially a JCT approach. When the project team's performance was reviewed at the end of the project, it was accepted that the NEC procedures had not been used, and everyone involved resolved to do better next time.

On the next NEC projects, effort was concentrated on applying the compensation event procedures correctly. The most dramatic effect was that deficiencies in works information became very apparent. This occurs because the pre-assessment of changes not only gives the customer the opportunity of making a decision in the knowledge of the cost and time consequences, but it also makes clear the cause of the change. In traditional practice many changes result from errors or gaps in works information. NEC makes these failures obvious and designers become more accountable to the customer for their own work in a way that does not happen with other contracts.

Consequently, the third major development in GDG's use of NEC was to concentrate on getting design information right. This was relatively easy to achieve as NEC forced consultants to recognise their own accountability. In parallel, it became clear that there was little point in having good works information if it was not matched with an equally clear statement of the cost and time implications. This led to the emphasis on using priced activity schedules to link design information to cost and time.

The fourth major development in the approach, which is the point currently reached, was recognising that risks are central to making decisions about contracts. Hence the approach provides a detailed schedule of risks for every contractual relationship.

The approach aims to provide a management framework within which cost can be driven down to a point where any further reductions would reduce the value created for the customer. This needs to be achieved without exposing anyone to more risks than are allowed for in their contract, so everyone can win.

Partnering

GDG sees their approach as analogous to partnering, because everyone benefits from greater certainty. Customers get new buildings that deliver real value for them, and consultants and contractors can manage their own work in accordance with a clear plan. This is a key principle in ensuring profitable work.

GDG's view is that in project management a good philosophy is more important than any contract. No matter how good a contract is, if it is used in an adversarial spirit it will provide very limited benefits. NEC provides an effective supporting tool for their approach to project management and partnering.

NEC training

GDG does not try to work only with firms they already know. They are more concerned to set up project teams that will deliver the best value for their customers and firmly believe that progress depends on new ideas. Consequently, their projects often include people with no experience of using NEC. They are required to attend training workshops, the learning from which is reinforced by coaching and guidance as projects progress. This thorough approach is expensive, but provides net benefits by helping ensure projects run well.

Picture acknowledgments

Courtesy of Greig-Stephenson Architects and Francesca Yorke

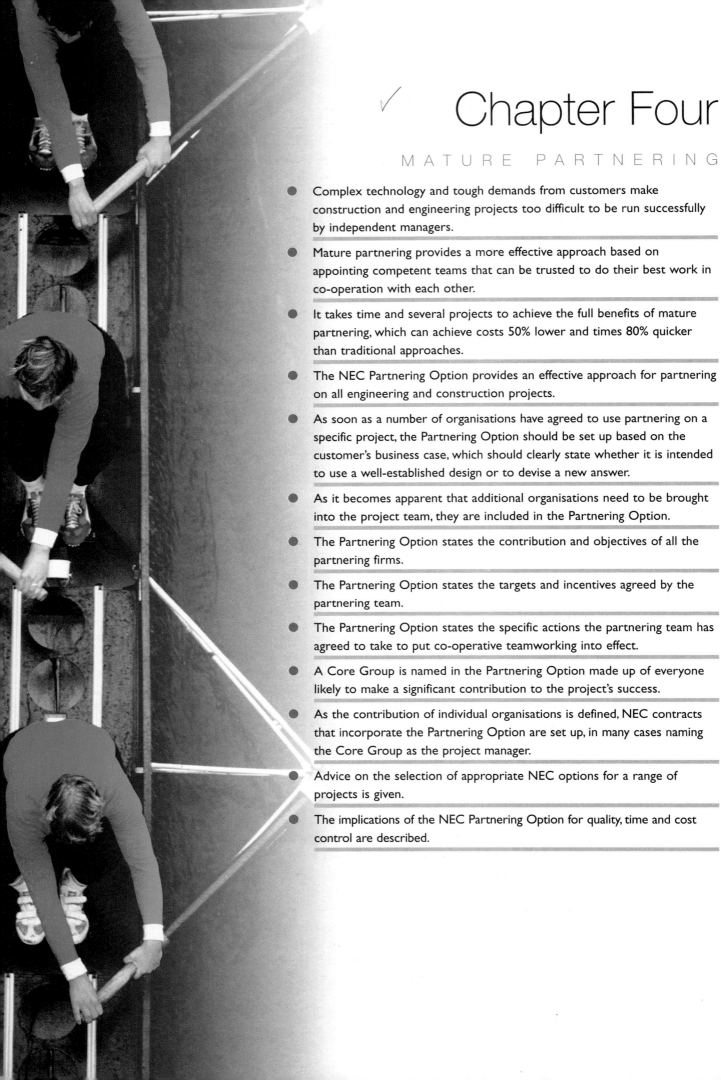

Chapter Four

- Complex technology and tough demands from customers make construction and engineering projects too difficult to be run successfully by independent managers.

- Mature partnering provides a more effective approach based on appointing competent teams that can be trusted to do their best work in co-operation with each other.

- It takes time and several projects to achieve the full benefits of mature partnering, which can achieve costs 50% lower and times 80% quicker than traditional approaches.

- The NEC Partnering Option provides an effective approach for partnering on all engineering and construction projects.

- As soon as a number of organisations have agreed to use partnering on a specific project, the Partnering Option should be set up based on the customer's business case, which should clearly state whether it is intended to use a well-established design or to devise a new answer.

- As it becomes apparent that additional organisations need to be brought into the project team, they are included in the Partnering Option.

- The Partnering Option states the contribution and objectives of all the partnering firms.

- The Partnering Option states the targets and incentives agreed by the partnering team.

- The Partnering Option states the specific actions the partnering team has agreed to take to put co-operative teamworking into effect.

- A Core Group is named in the Partnering Option made up of everyone likely to make a significant contribution to the project's success.

- As the contribution of individual organisations is defined, NEC contracts that incorporate the Partnering Option are set up, in many cases naming the Core Group as the project manager.

- Advice on the selection of appropriate NEC options for a range of projects is given.

- The implications of the NEC Partnering Option for quality, time and cost control are described.

Teamworking and networks

Mature partnering of the kind described in The Seven Pillars of Partnering depends on a new approach to construction and engineering projects in which key decisions are made by interdependent teams. For mature partnering to be successful, the people making up the teams have to be technically competent and have well-developed project management skills so that they understand, for example, how to work to programmes and budgets.

The teams required for a project work co-operatively, both within individual teams and in interactions between teams. The work of all the teams is co-ordinated by objectives and targets, agreed on the basis of discussion and consensus. Teams are guided by feedback so that decisions are made as an integral part of the direct work, not as a separate management activity. In a sense, well-developed networks develop the capacity to self-organise, dispensing with the need for a separate management hierarchy.

This fundamental change has become necessary because of rapid changes in technology and new demands by customers.

Technological complexity

Individual technologies have become more sophisticated, and decisions about them now require teams rather than individuals. This is now widely understood, and hence the wide interest in team-building activities throughout the construction and engineering industries. However, an even more important change is that the design, manufacture and assembly of modern engineering and construction facilities require many different technologies. As a result, large processing plants and modern buildings give rise to the most sophisticated and complex projects in which the work of many separate teams has to be co-ordinated.

Demanding customers

The second fundamental change facing the construction and engineering industries is that customers have become much more demanding. International competition has forced them to look for massive improvements in efficiency. So they question every aspect of their businesses, including their investments in construction and engineering work. As a direct result, the industry faces new demands for faster completions, lower prices, more reliable quality and greater certainty. As The Seven Pillars of Partnering describes, major customers have taken the initiative in helping construction and engineering firms meet these new demands. The Egan Report's recommendations flow directly from the experience of these leading customers.

Management reaction

Technological complexity initially led experienced project managers to concentrate on establishing milestones and providing sufficient time and resources to give teams a reasonable chance of meeting them. This approach is inevitably slow and expensive, but is the best answer so long as managers stick to the idea of independent work and individual liabilities. However, complex projects that are not well managed suffer from much higher levels of time and cost. This arises by default as projects get out of control and over-run completion dates and contract sums. This is broadly the situation described in the Latham Report, but its recommendations have now been overtaken by the new and insistent demands from customers for significantly better performance. As a result, the industry is now given less time and resources to deal with increasingly complex projects.

The combination of increased technological complexity and more demanding customers has made construction and engineering projects too difficult to be run successfully by independent managers. Plans for fast, efficient work based on independent activities fail. Inevitably, some teams miss deadlines and the knock-on effects cause delays, wasted effort and demotivation for others.

Emergence of networks

The approach that has emerged to deal with modern complex projects is to appoint competent teams and trust them to co-operate in doing their best work on the basis of discussion and consensus; and to accept that the resulting network of teams is responsible for all the outcomes. Where this approach is well developed, complex projects are carried out more efficiently and quicker than with management-based approaches that see projects

Teamwork

Teamwork is needed on engineering and construction projects because the vast majority of modern customers need answers quickly, so there is not enough time to wait for individuals to do their work independently. Work has to be closely integrated so it will be finished quickly. The pressure to work interdependently in teams is even greater on projects where new answers are needed. Work inevitably passes back and forth between the various specialists involved in searching for new answers. As a result, the required work cannot be predicted very far ahead with any great certainty and it is not possible to produce works information that tells people what they will be expected to do in sufficient detail for them to commit themselves to fixed costs or times. Tasks are discovered as the project progresses, new decisions are made and costs and times can be determined.

When work is interdependent, needed quickly and subject to uncertainty, the highest levels of efficiency come from people working together in teams. There is a mass of research evidence showing that for tasks as complex, uncertain and fast as modern engineering and construction projects, teamwork delivers better results than individuals working independently. This means that contracts should require people to work together, accepting joint responsibility for the outcomes.

Teamwork builds on the strengths and talents of different individuals so that their joint output is greater than the sum of their individual efforts could ever produce. Different points of view are respected and provide the inspiration for new ideas. In effective teams, people compensate for each other's weaknesses. It is normal for them to help any members falling short of the required performance. If things go wrong, time and resources are not wasted in allocating blame. Instead, every effort is concentrated on finding the best solution. This means that team members do what is needed, not what a predetermined contract tells them to do.

Research into effective teamwork shows that it depends on a number of distinct roles being played. The most widely used classification of these essential roles is that developed by Belbin, which comprises the following eight roles:

1   Co-ordinator: the natural chairperson of the team. Co-ordinators are good at clarifying goals and ensuring that the team agrees on priorities and reaches decisions.

2   Shaper: presses for action, finds ways around obstacles and drives the team to action.

3   Plant: the creative source of original ideas and solutions to difficult problems.

4   Monitor–Evaluator: carefully dissects ideas that the team is considering, weighs up the options and identifies problems.

5   Resource Investigator: the extrovert, enthusiastic, inspiring communicator who develops the external links that bring new contacts, ideas and developments into the team.

6   Implementor: the practical organiser who turns ideas into manageable tasks and then schedules and plans them.

7   Team Worker: holds the team together by supporting others, by listening, encouraging and understanding.

8   Finisher: checks details, worries about schedules and chases others with a sense of urgency that helps the team meet its deadlines.

Belbin's research suggests that all eight roles are needed to form effective teams. One person may play several roles but all need to be provided by someone if teams are to be effective and achieve their best work. Subsequent research has identified a ninth role, that of the Specialist. This recognises the fundamental need for technological competence if teams are to be effective in adding value for customers.

Many modern engineering and construction projects are too large and require too many different technologies to be undertaken by just one team. Therefore the work has to be subdivided into team-sized tasks. This commonly gives rise to various types of teams, including the following:

● Basic work teams undertake direct design, manufacture and construction work.

● Technology cluster teams establish design and management frameworks for each major element of the project, e.g. foundations, structure, external envelope, internal divisions, services, finishings and external works.

● Project core groups establish the overall design and management framework for the project. They include everyone likely to make a significant contribution to the project's overall success.

● Task forces find the best answer to difficult problems within a predetermined time-scale. They bring together experts and those project members with a close interest in solving the problem.

Each of these teams needs to be technologically competent, include all the essential team roles, and be skilled in designing its own methods of working, deciding which teams it should interact with and finding ways of improving its own performance. Given this broad competence, research shows that the most efficient approach is to empower the teams to make their own decisions so that they form a self-organising network guided by the currently agreed project objectives. There is no hierarchical management structure, simply a flexible, adaptive system responding to feedback. In effect this is teamwork within a higher level organisation comprising many teams.

Why is co-operation efficient?

Costs in engineering and construction projects arise from undertaking the direct work, interactions between teams doing that work and from overhead provisions needed to support the direct work and interactions. All three kinds of costs are influenced by project teams adopting co-operative ways of working.

Direct work

Co-operation influences the costs of any given direct work by making it more likely that the team undertaking it will have accurate information about what it is required to do and by reducing the chances that its work will be interrupted by other project team members. These benefits come from the greater understanding that team members have of each other's needs as they use workshops and other face-to-face meetings to discuss their individual interests. As this co-operation develops into long-term partnering, so the benefits grow even more as the work of teams is based on well-established designs with which they are very familiar. Research shows that the resulting learning curves quickly deliver cost reductions of 20% or more, and these reductions grow larger with more repetitions of similar patterns of work.

Interactions

Co-operation influences the costs of interactions by allowing teams to eliminate all those not essential to their direct work. It takes considerable experience of co-operating to eliminate all the interactions that are primarily designed to defend individual positions in any future disputes, or to establish the basis for a claim. As co-operation takes root, so defensive communications reduce and those that remain are purposeful and direct. Research shows dramatic reductions in the number of interactions over series of similar projects where co-operative methods have been adopted. For example, the number of written communications has been found to fall by more than 90% over four years.

Overheads

Co-operation influences the costs of overheads by making many traditional activities redundant. Where co-operation turns into long-term partnering the benefits can be very significant. There is less need to make provisions for dealing with claims and disputes, and there is often less need for competitive tenders and marketing designed to get onto tender lists. In general, more decisions can be made by direct work teams and so far fewer issues need be referred to head office. These benefits typically reduce overhead costs by at least 50% over several projects in which teams have worked co-operatively.

Cost effects

Research suggests that the combined effects of these benefits reduce project costs by up to 50%. The following table is consistent with a variety of research data and shows how the separate effects combine:

	Direct costs	Interaction costs	Overhead costs	Total costs
Traditional methods	50	40	10	100
Early co-operation	45	25	10	80
Developed co-operation	38	15	7	60
Mature partnering	35	10	5	50

These research results compare like with like, so both projects based on well-developed designs and those using individual designs that use mature partnering tend to cost 50% less than traditional projects which provide similar value for their customers.

However, in both the traditional and mature partnering approaches, individually designed projects are more expensive than those using well-developed designs meeting similar function and quality criteria. It is of course to be hoped that with either approach individually designed projects provide customers with sufficiently greater value to compensate for their higher costs.

as a series of independent activities for which individuals can be made liable.

Interdependent teams

Projects are no longer seen as principally comprising two independent parties – the employer represented by the project manager, and the contractor – that have different interests. Professional consultants and subcontractors are no longer seen as essentially supporting the interests of the two main parties, required, for example, to communicate only through them.

The new approach is based on bringing key people into teams that take joint responsibility for undertaking projects in ways that take account of all their various interests. Research in many industries, including the UK engineering and construction industries, shows that this is far more efficient than traditional management-based approaches, which fragment project teams by concentrating on defining independent responsibilities and liabilities.

Co-operative teamwork

The speed demanded by modern customers means that project teams have to be assembled rapidly and start work fast. All the key players need to be in place as early as possible. These often include, in addition to the customer's representatives, financiers, urban and regional planners, design consultants, main contractors and specialist contractors. In practice this means that many distinct teams work together on the basis of co-operative teamwork.

Teamwork brings key individuals together to make decisions by discussion and consensus. Problems are identified early so that everyone can adjust their activities to actual progress. It is common during the most intensive stages of creative work for all the key members of the team to work in a common office. For especially demanding projects, such offices have a role to play throughout most stages of the work. Communication has been found to be fostered to a high degree in project offices, so decisions are based on a true picture about the real situation facing the project.

Partnering encourages these good effects because it helps team members co-operate in searching for the best possible answers for their mutual benefit. As a result, the vast majority of modern engineering and construction work is undertaken by teams, and the most effective of them use partnering. The common outcomes, as described in The Seven Pillars of Partnering, are that effort is concentrated on effective work that delivers good value for the customer and good profits for the engineering and construction firms involved, costs are low, quality is good and projects are completed quickly and on time.

Developing efficiency

Once a project team has worked together on several projects, it can concentrate on achieving higher levels of efficiency by co-operating on undertaking only those activities that directly add value for the customer. The Egan Report describes this as the philosophy of lean production. This means eliminating activities that consume time and resources but make no direct contribution to producing the required facility and support services. There is often scope for improving direct design and construction activities, but the biggest improvements come from streamlining communications. There is massive scope for simplifying and eliminating paperwork, electronic communications, meetings and all the other forms of interaction used in traditional management-based practice. This applies to at least some of the NEC procedures; e.g. most of those surrounding compensation events add no value for the customer but consume time and resources. They provide real benefits as people move away from traditional approaches, by ensuring that everyone undertakes their roles and responsibilities co-operatively in the interests of the project. When these good habits are established, the formal procedures can safely be streamlined.

Modern communication

A recognition of the waste arising from formal communications in management-based approaches has coincided with developments in communication technology that make it possible for people to work interactively from remote locations. Much routine communication is best carried out this way because digital technologies dramatically reduce communication costs. This is the case for much of the communication required within well-developed supply chains. Digital technologies can also be used for most communication in project teams that use established patterns of construction technologies. This is because most interface problems have been identified and solved on earlier projects. In such situations communication becomes largely a matter of simple exchanges of information about progress, and is much faster than traditional, procedure-bound methods.

This ideal is as yet fairly unusual in construction and engineering projects, and there remain many situations where it is more efficient for people to work face to face. This is because creative work, tackling difficult problems, resolving disputes and similar situations benefit from using all the human senses in trying to understand each others' point of view, and so feeling confident enough to co-operate in finding the best answers. Hence the continuing use of well-structured meetings, workshops, project offices and social events.

Project types

The significance of the new ways of working is that project teams now have to make decisions about which communication methods to use for which projects and for which parts of projects. In very broad terms, this classifies engineering and construction work into two broad categories. The first type of work uses well-established answers that can be undertaken using virtually automatic forms of interaction requiring very little face-to-face communication. These should provide the majority of the industry's work. The second involves finding new answers and needs a great deal of face-to-face communication. Teams undertaking these individually designed projects often use modern information technologies to access wide ranging feedback and databases in their search for the best possible answers, but their principal means of communication are the various forms of face-to-face meeting.

Both of these broad categories of projects may be undertaken for customers who have established long-term relationships with consultants, contractors and suppliers. Equally, both may be undertaken for occasional customers of the construction industry. The way in which these two types of customers should set up their project teams is well described in the Construction Industry Council's guide to project team partnering, under the headings of Alliances and One-off projects.[1]

For both types of customers, most projects involve to some extent both well established and new answers. So projects that use well-developed designs predominantly rely on well-established interactions using the latest communication technologies, and use face-to-face methods mainly to deal with problems. Individually designed projects involve a great deal of creative work and so use face-to-face interactions extensively, supported by considerable use of communication technologies to access information.

1 Construction Industry Council, A Guide to Project Team Partnering, London, 2000.

When project teams decide to use established design and construction methods for which the costs are known, then mature partnering enables them to work quickly and efficiently. These benefits influence the direct design and construction work, but the biggest benefits come from streamlining communications between teams used to working together on familiar tasks. Necessary communications are increasingly handled automatically using sophisticated information technology. The costs associated with these interactions can be close to zero as the direct work fits together almost instinctively.

Inevitably, projects that require teams to find new answers result in higher costs and longer timeframes. Even in these cases mature partnering provides advantages compared with traditional approaches by using teamworking to make communication faster and more effective. This in turn enables teams to consider more options and so find better answers.

The NEC Partnering Option

Recognition of the benefits of mature partnering led to the NEC Partnering Option, which establishes a common framework to guide project teams in co-operative teamworking. A completed example of the NEC Partnering Option is included in Chapter 6. The Partnering Option is given effect by incorporating it in individual NEC contracts between members of the project team. The partnering team is established by incorporating the same Partnering Option in the individual contracts of all those included in the partnering arrangement.

This arrangement allows contributors who have well-defined roles that are unlikely to have a significant effect on the overall success of a project to be employed under NEC contracts that do not include the Partnering Option.

The formal status of the Partnering Option is established by making it part of individual contracts. Nevertheless, in deciding on its use, it is helpful to think about the Partnering Option as the essential agreement between the partnering organisations expressed in straightforward terms. Then the individual contracts can usefully be regarded as providing more detail.

The NEC Partnering Option provides an effective approach for all engineering and construction work, including individual projects, programmes of projects and the long-term development of better ways of working. The Partnering Option can be used for all these situations, some of which may need the support of individual contracts only after considerable progress has been made. Indeed, it may well be that individual contracts are set up only when a specific engineering or construction project begins, because the Partnering Option provides a sufficient basis for work prior to that stage. Where this approach is adopted, some provision will be needed for payment for work carried out prior to the formal contract being set up.[2] This is consistent with the Egan Report's recommendation that the industry should learn to work without contracts.

The Partnering Option will be set up when a number of organisations have agreed that they will partner. The *Client* named in the Partnering Option is the customer for whatever is to be produced, and the *Client*'s objective is best stated in terms of the customer's business case. This will be as specific or general as the situation allows. In some cases customers know in detail what they want, can describe it specifically and know what they want to pay for it and when it must be complete. They may, for example, say they want a warehouse of a specific well-established type and specified capacity, on a site they already own, for £12 million by 1 April 2001. Other customers have less understanding of their specific needs and their objective will be couched in general terms. They may, for example, say they want to re-house 2000 designers and engineers in modern accommodation that supports the creative use of information technology and other design tools on a given site for £30 million by 1 September 2001.

In even more general terms, they may say they want to improve the quality, time, cost, certainty of delivery and flexibility in use of warehouses by specific percentages over the next five years. In this case, the client could be a retail business that owns a major distribution system, a developer specialising in distribution warehouses, or a contractor who plans to market a range of state-of-the-art warehouses. The key point is that the Partnering Option caters for all these very different situations by allowing the partners to describe their agreement in their own terms.

---

2 *A simple purchase order could be used.*

The Schedule of Partners lists the organisations brought together to achieve the client's objectives. The nature of their contribution will be described in a similar level of detail to that employed in the client's objectives. Importantly, the Schedule of Partners states each partner's objective. In most cases this will include making a specified profit and covering all their costs. It may also refer to staff development, aims for the engineering or construction outputs, improvements in productivity, safety, quality, speed and certainty of delivery, and such things as zero defects. These may simply mirror the client's objectives, but may go further. It is best to keep the statements of objectives short by stating the partner's major objectives that add to the client's objectives.. However, it is important to include each partner's main commercial objectives.

It is helpful to have all the individual objectives recorded so the partnering team members can take account of each other's interests in making decisions. However, their mutual objectives, which may provide more than their individual objectives or may require some compromises, are stated as Key Performance Indicator Targets. These are linked to incentives stated in terms of payments to each partner if Key Performance Indicator Targets are improved upon or achieved.

The list of Key Performance Indicator Targets and incentives will normally result from a partnering workshop where all the partners discuss their individual objectives and search for mutual objectives that give everyone the best possible deal. Common Key Performance Indicator Targets are the actual cost related to the client's budget, and the required completion date.

The incentive payments to each partner need not and normally will not be equal. They may be in proportion to the estimated cost of the partners' contributions, their ability to influence the success of the work, or some other basis.

The Partnering Information describes how the partners have agreed to work together in a spirit of mutual trust and co-operation. This is likely to have been decided at the same workshop as agreed the Key Performance Indicator Targets and incentives. It normally covers such matters as: the way the Core Group will work, including arrangements for chairing meetings;[3] the composition and work of a steering group and any task forces the Core Group sets up; and partnering workshops, including team-building, value engineering, value management, risk management and common information systems. The case studies in Chapter 5 provide examples of the approaches used by a number of experienced partnering teams.

The members of the Core Group are stated in the Schedule of Core Group members. The Core Group is crucial to the success of the partnering arrangement. It should have few members so that it can work as a genuine team, but it should include everyone likely to make a significant contribution to the success of the joint work. The members should all be empowered to make decisions on behalf of the organisations they represent. It should include the client's representative who is empowered to commit the customer. This is often the person responsible for the budget in the client's organisation. It usually helps to have a representative of the end users, who can speak with authority on the way the end product will be used and operated. The main design and construction organisations should be represented. The people responsible for overall control of quality, time and cost should be members of the Core Group. Any specialist knowledge central to the success of the work should also be represented for as long as this knowledge is central to the Core Group's decisions. This may be for only part of the time for which the work is underway, and the Partnering Option provides for individual joining and leaving dates to be stated.

The NEC Partnering Option provides a link between the individual NEC contracts that bring firms into the project team. The individual contracts have to be selected and set up carefully, taking account of the nature of the particular project to ensure that they do not give rise to conflicts of interest within the partnering team. One approach that helps avoid problems is to name the Core Group as the Project Manager in the individual ECC contracts. This recognises that partnering relies on teamworking. Obviously, the project management functions have to be carried out by individuals, but it is more efficient to make this an integral part of their direct work. This is usefully illustrated by describing how the main functions of quality, time and cost control are effected.

---

3 Although the NEC Partnering Option provides for the Client to convene and chair meetings of the Core Group, in practice it is often most effective for one of the other Partners to undertake these roles. This is illustrated in the case studies in Chapter 5. Such situations can be accommodated by agreement between the Client and the other Partners, and stated in the Partnering Information as the Option conditions require.

Quality control

Best practice is for quality control to be an integral part of everyone's work. The NEC approach of appointing an independent supervisor provides a good starting point. The supervisor should approve the quality control systems used by each of the firms involved in the project. These should include detailed schedules of the tests to be applied to each of the elements and systems in the new facility. The schedules should form part of the ECC contract's works information, so that the ECC supervisor's role of checking that the works are constructed in accordance with the contract becomes a process of monitoring the test results. Mature practice relies on self-certification by the contractor, so the supervisor's role becomes one of auditing the paperwork that sits behind the certificates and making spot checks on the finished work. The logical extension of this, as the case studies illustrate, is for the contractor to be named as the *Supervisor* in the ECC contracts.

Whichever of these arrangements is adopted, it is important that the Core Group members work together co-operatively to ensure high standards of quality. This means that the new facility is handed over to the customer fully complete, tested, commissioned and with zero defects. Anything less is bad practice that undermines the industry's reputation.

Time control

The timetable for the Core Group's work provides the highest level of programme for the required work. The programmes required by individual contracts provide more detailed programmes for the individual contributions in a hierarchical relationship with the Core Group's timetable. The individual programmes are co-ordinated by the Core Group. In practice, the day-to-day co-ordination will be the responsibility of one member of the Core Group. Often this is the contractor's project manager, but it could be the customer's representative or some other member of the Core Group with specialist programme management skills.

The timetable, like the supporting individual programmes, will be as detailed as the current definition of the work allows. This means that initially on one-off projects it is most likely to be in outline form, because at the time when the Partnering Option is set up, the nature of the work has not been determined. In such cases the initial timetable comprises little more than key milestones that must be met to achieve the completion date required by the customer. As decisions are made, more detailed milestones and broad method statements can be added.

Then, as yet more decisions are made, the detailed programmes and method statements required by individual contracts are added within the overall structure of milestones. The partnering arrangements should ensure that the common aim of everyone in the team is to do whatever is necessary to finish on time. NEC procedures for agreeing what is to be done and then sticking to what is agreed provide the kind of discipline needed to control time, as is well illustrated by the case studies.

Cost control

The Core Group undertakes cost control by establishing and maintaining an up-to-date cost plan based on the overall financial framework in the NEC Partnering Option's Key Performance Indicator Targets and incentives.

A key issue in establishing the overall financial framework is deciding who takes the financial risk that the project budget may be exceeded. Arrangements where it is shared tend to undermine partnering attitudes. It is more effective for the main financial risk to be carried either by the client or by the contractor. It makes sense for whoever carries this risk to be named in the Partnering Option as the *Client*.

In general, when new designs are required, it is best for the client to carry the main financial risk, and where well-established designs can be used it is best for the contractor to carry the main financial risk. This simplicity allows everyone involved to concentrate on ensuring that cost is controlled within the overall budget, without being distracted by worrying about their own financial situation. Some clients and some contractors are reluctant to accept this approach, and so share the risk. In some cases this means that costs are predetermined and treated as prices. The NEC provides for all these situations, so that whatever is decided the individual projects can use the appropriate NEC option for dealing with the money.

The NEC provides for detailed cost control, so any threats to a cost target and any opportunities to make savings are identified early by the procedures in individual contracts. One of the Core Group's most important tasks is to co-ordinate the cost control information resulting from individual contracts. This co-ordination will be undertaken on behalf of the Core Group by one member, sometimes the customer's representative, a consultant cost manager or the contractor's project manager. The Core Group exercises overall cost control, using the current cost plan to make decisions about cost threats and potential savings.

Individual contracts

The arrangements described above can be applied to all types of projects, but the use of individual NEC contracts with an NEC Partnering Option will depend on the nature and circumstances of the engineering and construction work. The following descriptions deal with broad categories of projects and suggest how NEC contracts can be used in various circumstances with an NEC Partnering Option.

Individually designed projects

Individually designed projects provide innovative solutions for customers who want a facility that is distinctive, uses new technologies, meets unusual performance requirements, or in some other way poses a unique challenge. Such projects are best tackled by a team of competent and experienced people willing to look creatively for the best new answers using partnering.

Best practice works on the basis that the client defines the business case for a new facility which describes the function and quality in broad terms and establishes the programme and budget. Then a project team is set up to find the best possible answer within these criteria. They use a mature partnering approach which

Cost control

Best practice cost control begins with the client's business case, which defines the function, quality, time and value required in a new facility. This establishes the criteria to be met by the project team and their overall budget. The feasibility of the criteria and budget is tested by reference to similar projects, with the aim of setting targets that require the team to make some defined improvements on previous best performance.

A cost plan is produced based on the required function, quality and time in as much detail as possible. It does not include any contingency allowance, so that the client will get full value for the agreed budget. The cost plan is progressively developed as further design decisions are made until there are well-defined targets for each element and system in the finished facility. Best practice is for each target to be agreed in co-operation with the firm responsible for the particular work. The targets should be based on everything going well, so they do not include contingency allowances.

When the cost plan is in a reasonably robust state, the project team should hold a workshop to discuss each of the targets, the assumptions on which they depend, the risks and any remaining uncertainties. Then the whole team should be required to formally accept joint responsibility for making the cost plan work, so the budget is achieved and the client gets everything needed from the project.

As agreements are reached with the firms responsible for each element or system, their part of the cost plan can be more detailed. As further decisions are made their cost effects are monitored and recorded once a week in a detailed cost report. This states the cost status of each element or system and highlights any threats to the target and any opportunities to make savings. Cost reports are reviewed by the project team weekly, and decisions made about all threats and opportunities, so that the project delivers the best possible value for the client within the budget.

When difficult cost problems arise it is the whole team's responsibility to search for savings to get the project back within budget. This often involves setting up a task force to find the best possible answer to some difficult problem. It may mean that some cost targets have to be cut and different designs produced. The most effective approaches do all this without reducing firms' profit margins or giving the client worse value. They challenge project teams to be creative in finding good answers.

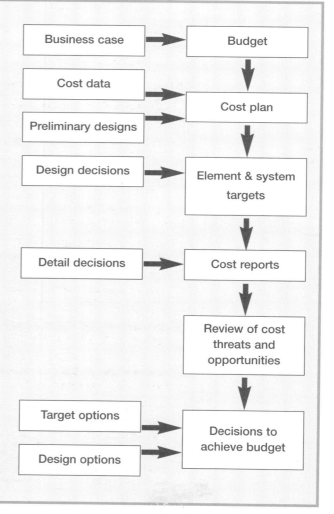

NEC has been proven to be effective for individually-designed landmark projects, such as the Eden Project and Cardiff's Millennium Stadium

Key features of mature partnering for individually designed projects

1   Talented people drawn from separate firms are brought together to produce an innovative new facility and support services tailored to a customer's specific needs.

2   Project teams include representatives of the customer's internal divisions and external bodies with an interest in the new development, as well as the key engineering and construction firms.

3   Extensive initial work is often put into exploring the potential benefits and costs of a new facility This may include study visits, research, workshops with experts, task forces set up to explore specific aspects of the proposals, and other creative techniques to ensure that many options are considered.

4   The customer's business case determines the function, quality, time and cost.

5   Contracts with customers either provide single point responsibility (design build) or bring the customer centrally into the project team (construction management).

6   Project teams use partnering in all their relationships to search for the best possible answers within the constraints set by the customer's business case. Key members of the team often work together in a common project office to facilitate communication.

7   The firms that provide members of the project team are guaranteed a predetermined profit plus all their properly incurred costs, determined on the basis of open book accounting backed by tough audit procedures.

8   Tough quality control systems are applied to every aspect of the work and clear records of test results are maintained and open to everyone involved.

9   The completion date required by the customer is treated as fixed, and time control provides firm milestones to ensure it is achieved. Between milestones work is planned flexibly to allow the team's search for the best possible answers to continue as long as possible without delaying completion.

10   The customer's budget is regarded as fixed and the whole team takes responsibility for ensuring it is achieved. Tough, detailed cost control is used to ensure that all cost threats and potential savings are identified, and the project team makes carefully considered decisions about all of them.

11   Firms specialising in individually designed projects work with users, neighbours and owners in actively researching the performance in use of the facilities they produce to ensure that lessons are learned. This is an important part of developing more effective ways of working creatively.

aims to provide the client with excellent value in a new facility that meets or exceeds all his requirements, guarantees all the engineering and construction firms involved a fair profit, and reimburses all their properly incurred costs. This essential deal is recorded in an NEC Partnering Option, which is incorporated in the individual contracts of the key partners.

There are two main ways in which NEC provides an appropriate contractual framework. In broad terms, these are a design build approach and a construction management approach.

Individually designed design build

Many clients like the single point responsibility that design build provides, and so assemble a team that includes a firm willing to take formal contractual responsibility for the whole project.

Client's representative

The client will normally appoint a representative to organise all the internal and external interests that have to be taken into account, and help select the main contractor. The client's representative may be a member of staff or a consultant, most probably appointed using an NEC Professional Services Contract using either Option C, Target Contract, or Option E, Time Based Contract, depending on how tightly the customer wants to limit expenditure. This will incorporate the initial version of the Partnering Option.

There may well be a need for advice from consultants or specialist contractors during the initial stage. Firms involved in this way can be appointed on a similar basis to that used for the client's representative.

In selecting a design build contractor, the requirements of individually designed work need to be kept clearly in mind. Various organisational arrangements are possible and the best might be, for example, a joint venture between an innovative architect, a multi-discipline engineering consultancy and a management contractor. There are many other possibilities, the key requirement being that the resulting organisation is able to take formal contractual responsibility for creative work.

Whatever decision is made, the project manager then works with the contractor in selecting the other members of the Core Group, which then undertakes the project manager's role defined by the ECC throughout the project.

Main contractor

The main contractor is responsible for the design and construction of the whole of the works as part of the Core Group. The choice of the most appropriate ECC option depends on the risks the customer is prepared to accept. The Partnering Option defines the overall deal, but some customers prefer to have a lump sum contract based on detailed works information, while others see advantage in paying the contractor's actual costs and concentrating on finding the best possible answer quickly.

ECC Option A, Priced Contract with Activity Schedule, based on works information that describes a specific design in sufficient detail for the contractor to be confident of completing the project within the contract price, provides the greatest cost certainty for the customer. Some contractors may be willing to rely on the Partnering Option prior to that stage, but many will prefer to have a PSC in place to ensure they are paid for their design work. In either case, establishing a firm price for a new design before work begins on site can be a very slow approach.

For customers who want a firm assurance about costs before authorising work, but also want an early completion, the NEC, Option E, Cost Reimbursable Contract provides the best approach. The design will need to be developed as distinct packages, each of which could be treated as a compensation event. The NEC procedures provide for contractors to provide quotations for compensation events which become firm once they are accepted. Under Option E the contractor is still paid Actual Cost plus Fee and the compensation event only adjusts the forecast total of the Prices. However, by adopting this approach, work progresses, in effect, as a series of firm price contracts.

Option C, Target Contract with Activity Schedule, provides the right approach for customers who want to pay the actual costs, because this allows the contractor to concentrate on finding the best answers, without having to worry about whether his price covers his costs. It is the approach that is most consistent with mature partnering and it works best when the works information is the customer's business case, including the project's programme and budget exactly as stated in the Partnering Option. The contractor's individual contract may provide for a zero share of cost savings or over-runs because the Partnering Option provides the incentives. The effect of these arrangements is to make the contractor responsible for working with the Core Group to produce the best possible answer to the customer's requirements within defined time and cost targets. The target (total Price of the activities in the activity schedule) will be changed only if the customer's business case needs to be altered. The forecast of the total Actual Cost will be developed as design information is produced. This mature approach depends on considerable experience of working together, so that there is a soundly based trust between the customer and industry firms.

Core Group

The Core Group includes all the people likely to make a significant contribution to the success of the project. It usually includes representatives of the customer's key internal and external interests, one of whom will be the client's representative named in the Partnering Option, key members of the main contractor's staff, and other team members as and when their work is central to progress. The Core Group is responsible for looking for good answers and, having agreed which are the best available, putting them into effect.

It often makes good sense to accommodate the Core Group in a project office so that communications are direct and there is less likelihood of misunderstandings or wasted effort.

The selected designers will have contracts with the contractor based on the PSC, and specialist contractors will have contracts with the contractor based on the ECS or ECSC. They will include the payment option that fits their particular contributions. Thus, where they are directly involved in the partnering team's search for the best answers, they are most likely to be paid on a cost reimbursable basis and their contracts will incorporate the Partnering Option. However, if their work can be predetermined, is unlikely to change and they are not closely involved in key decisions, they are more likely to be paid on the basis of a priced contract with activity schedule, and they will not be brought into the partnering arrangement. Their work can be taken into account with confidence in the time and cost control systems. Should it be necessary to make changes, the NEC compensation event procedures serve to make carefully considered decisions at an early stage so that time and cost control can be maintained.

Individually designed construction management

The organisational flexibility provided by construction management provides the most natural approach for individually designed projects when the customer wants to be closely involved in the work.

Lead designer and construction manager

At the start of the project, the customer appoints two firms, one to be the lead designer and the other to be the construction manager, to work with the customer's representative in undertaking the project. They are all named in the NEC Partnering Option. The lead designer and construction manager are appointed on NEC Professional Services Contract using Option E, Time Based Contracts, which incorporate the Partnering Option. This means they are paid for all the work involved in finding the best answer for the customer, and so do not have to worry about their own financial position.

Some customers use NEC Professional Services Contract, Option C, Target Contract, in the belief that it provides useful incentives. However, since the NEC Partnering Option provides incentives for those involved in the partnering arrangement, the target contract provides no additional benefits, but does require extra work in maintaining the target in line with compensation events.

Flexible project team

Other firms are brought into the project team as and when they are needed. These typically include various types of professional designers and specialist contractors. They are appointed either under the PSC or the ECC. Any one of the payment options provided by the NEC may be appropriate for each contract, which will incorporate the Partnering Option for key members of the team. Those most centrally involved in the project's success will also become members of the Core Group. Considerable effect could be given to their role by naming the Core Group as the project manager in the other ECC contracts.

The composition of each of these teams changes as decisions are made and the nature of the work moves from defining the customer's needs, to design, manufacture and then construction. The lead designer and construction manager, in discussion with the customer, agree a design and management framework that enables the other firms to play their full role, to be creative and to join in the search for innovative answers, yet ensures that their work is co-ordinated and in total meets all the customer's requirements.

Projects using well-developed designs

Many projects do not need one-off, new stream answers, but use well-developed designs supported by teams fully experienced in the required work. The most highly developed form of projects using well-developed designs are close to providing brand-named engineering and construction products. These are well-developed facilities that can be tailored to customers' specific needs. The physical facilities are supported by comprehensive packages of services that include helping to find the necessary finance and land, studies of how a new facility can benefit the customer, management of the new facility, maintenance and training for the customer's staff.

The emergence of these well-developed products is a comparatively new development that has been given considerable impetus by the private finance initiatives of governments around the world. The resulting opportunities to build new businesses have encouraged alliances of developers, designers, contractors and facilities managers to develop products they are prepared to produce and operate on a long-term basis.

Engineering and construction projects that use well-developed designs provide consistently high levels of efficiency. Modern manufacturing technologies allow this efficiency to be combined with stylish and attractive designs that make a positive contribution to the health, happiness and economic well-being of users and owners. Such approaches require designers to work within a discipline that reflects the limits of current technologies. However, this does not result in narrowly standardised answers, because modern manufacturing provides more than sufficient variety and flexibility to allow great design to flourish, as is amply demonstrated by the work of the UK's leading architects.

Contractors for projects using well-developed designs

Highly developed products are produced and marketed by substantial organisations. In NEC terms these organisations are contractors that are either multi-firm organisations using strategic partnering, or large well-established firms. They offer potential customers a choice of new facilities. The options are supported by marketing information designed to give customers reliable information about all aspects of the facilities and services on offer. Where the contractor is a multi-firm organisation it may well be sensible for them to set up an NEC Partnering Option to record how they intend to work together.

Selecting the contractor

Many experienced customers have well-established agreements with contractors who specialise in the type of facilities they use. When a new need arises, they simply order another new facility from their regular contractor.

Other customers may well consider proposals from several contractors in order to be assured that they are getting good value. This may be done through formal competitive tenders, but more likely will involve interviews and negotiations. The ideal arrangement is for the industry to encourage independent evaluations of competing products to be published. Then customers can make informed judgements without all the expense of tenders, interviews and negotiations. It should be a high priority for the engineering and construction industries to ensure that high-quality, independent customer information is widely available. The internet now makes this easy to publish widely. This will make it much easier for customers to decide to invest in new facilities, because they can be confident about what represents good value and can make tough but realistic demands on the industry. However, even with good information available, some customers will seek specialist advice to ensure that they understand the descriptions provided by contractors, the risks involved and whether what is on offer really does represent good value.

Client's representative

Clients may decide to appoint a representative to help determine what is required and help select the best contractor. The most appropriate NEC option in the PSC for this appointment is likely to be Option A, Priced Contract with Activity Schedule. It may include an NEC Partnering Option, but producing this may be delayed until a contractor is selected.

An alternative approach is for the client to set up a team comprising representatives of all the internal and external interests affected by the project, and let them deal with the initial decisions. For experienced clients, this stage can be very quick, but where more decisions and investigations are needed the customer's team may work for several months before deciding on the best answer. Where external experts are involved they can be appointed using an NEC Professional Services Contract using either Option C, Target Contract, or Option E, Time Based Contract, incorporating an NEC Partnering Option.

Once the client's team have agreed the facility and services to be provided, selected the contractor and settled the terms, clients are likely to appoint one of the team to be their representative and provide a focus in looking after their interests. Here again, when an external consultant undertakes this role, an NEC Professional Services Contract using Option A, Priced Contract with Activity Schedule, incorporating the Partnering Option, is likely to provide a good basis for the appointment.

Main contract documentation

When agreement is reached between the customer and contractor, it can be recorded in an NEC Partnering Option which, where a multi-firm organisation is selected, may well be based on the Partnering Option used by the firms that make up that organisation. This brings all the key parties into a team with agreed mutual objectives. In addition, most customers for projects that use well-developed designs will want the detailed and specific terms of the agreement with the main contractor recorded in a formal contract. This should be straightforward and the ECC provides a sensible basis.

The first step in establishing an NEC contract is producing the contract documentation. On projects that use well-developed designs, the chosen contractor should be able to produce the contractual description of the new facility and any supporting services from their standard information. This should provide works information in the form envisaged by the NEC. The contractor should also have carried out a thorough site investigation as part of the negotiations leading to the agreement with the customer, which provides a robust basis for the site information required by the NEC. This complete package of information will be checked and agreed by the client's representative in the same manner as it would be if produced by consultants.

As far as the formal contract for projects using well-developed designs is concerned, the NEC option in which the contractor designs and constructs the whole of the works is the most suitable. This is likely to be used with ECC, Option A, Priced Contract with Activity Schedule, incorporating the Partnering Option. The completion dates for each activity are stated on the programme submitted for acceptance, so the customer can organise payments when they are due. There is no point in the programme being any more detailed than this unless it is expected that compensation events may arise. In the latter case, a well-organised contractor should be able to produce a detailed programme, and the data needed to establish changes to the Prices using Actual Cost plus the Fee envisaged by the NEC, from its own standard documentation.

The NEC approach to compensation events provides an appropriate framework for projects using well-developed designs, but in practice should rarely need to be used. Indeed, compensation events should arise only if the customer fundamentally changes his requirements. This is exactly what NEC provides, because changes made by the contractor to meet the customer's requirements are not compensation events. Beyond this, experienced contractors should absorb minor changes by customers as part of a policy of building goodwill, and only seek to change the price for major changes. The contractor may well need to deal differently with subcontractors and pay them for changes that may be minor in the context of the whole project, but significant in terms of the subcontractor's work. As part of the same policy of providing certainty for customers, it should be very unusual for a project using a well-developed design to be handed over even one day late. As far as quality is concerned, most experienced customers who have established a long-term relationship with the contractor will rely on the contractor's reputation, possibly backed by long-term insurance, to provide quality assurance and ensure that defects are dealt with quickly and properly. Where this is not the case, the customer is more likely to use an NEC approach to quality assurance.

Management by the contractor

Management of the project will follow the contractor's well-established approach. This does not change the project manager's role as described in the contract if the Core Group is named as *Project Manager*. It does mean that little of the Core Group's work will be dictated by the contract procedures. There will be very few or no compensation events. Construction will be carried out as a well-developed process in which all the milestones are met, so that payments are made for completed activities in accordance with the predetermined activity schedule and programme submitted for acceptance. The NEC Partnering Option brings the customer's representative into the management process to whatever extent is necessary to ensure that the customer gets everything promised in the original deal.

The NEC's role in the contractor's organisation

In mature partnering on projects that use well-developed designs, the separate firms that form the contractor's overall organisation work together on the basis of long-term strategic partnering. The NEC may well be used to provide a contractual framework between these firms of designers, manufacturers, contractors and specialists to provide a contractual safety net, should events go wrong or relationships break down. The procedures set out in the NEC are

> ## Key features of mature partnering for projects using well-developed designs
>
> 1 Substantial contractors market a range of products and packages of support services to categories of customers.
> 2 They publish detailed customer information that includes independently produced benchmarks of the value provided by their products and services.
> 3 For potential new projects, they undertake extensive studies with the customer, neighbours, local authorities and special interest groups to develop a wide understanding of the likely benefits and costs.
> 4 Contracts with customers are straightforward and guarantee performance in use, fixed prices, completion on time and zero defects.
> 5 Contractors have highly developed and efficient methods of working based on well-established supply chains using partnering in all significant interactions.
> 6 They provide extensive after-sales care in support of the facilities they produce to ensure that users, neighbours and owners get the best possible value.
> 7 They undertake market research and use feedback to search for ways of continuously improving their products and services.

more formal and detailed than those needed in well-established organisations where most activities have become virtually automatic and do not need to be documented. When such well-established procedures are in place, people simply work together efficiently in an almost instinctive manner and there is little paperwork.

The NEC Partnering Option provides exactly the right level of detail for a successful partnering arrangement with the contractor being named as the *Client*. Indeed, as suggested above, the Partnering Option describing the agreement inside the contractor's organisation may well be used as the basis of Partnering Options incorporated in individual contracts between the contractor and customers.

However, established answers cannot deal with every situation, and inevitably there are unique aspects of all projects, even if they concern only the site and its immediate surroundings. Established contractor organisations based on using well-developed designs have well-developed answers to most site-based issues, and will have experience of most situations that arise. This means they can accept the contractual responsibilities of providing customers with complete facilities at fixed prices, and meet firm completion dates and quality standards. The benefit of using NEC to provide a contractual framework is that, if completely unanticipated problems arise or the standard approaches do not work, there are procedures in place to deal with any situation that arises, including, if necessary, resort to an adjudicator.

Therefore, the main role of the NEC inside contractors' organisations using well-developed designs is to provide the comfort of having a formal contract that does not contradict efficient ways of working. Once into the habit of NEC's procedures and time discipline, the contract will rarely be referred to by any of the parties involved, but its existence helps give them all the confidence to concentrate on doing their best possible work on all the projects they undertake together. It is most likely that the various consultants and specialist contractors will have contracts with the contractor based on the PSC, ECS or ECSC. These will include the payment option that fits their particular contributions, and each contract will incorporate the Partnering Option.

✓ Chapter Five

CASE STUDIES OF NEC AND PARTNERING

- The Environment Agency's use of NEC as a partnering tool.

- Costain Limited's experience of NEC and Partnering.

- The Halcrow Group's approach to NEC and partnering.

The Environment Agency's use of NEC as a partnering tool

- The Environment Agency have an explicit strategy based on NEC and partnering aimed at improving performance by 20%.

- Consultants and contractors are carefully selected on the basis of competence and partnering attitudes.

- Partnering is put into effect through workshops.

- The Agency's experience shows that NEC supports partnering attitudes.

- Training in NEC procedures and project management is necessary.

- NEC causes project managers to think very carefully about their objectives before they prepare Contract Data.

- The Agency's strategy encourages contractors to look for ways of reducing costs and times.

- Individual incentives in target contracts need to be carefully aligned so that firms are not penalised when they incur extra costs to provide savings for others in the project team.

- Traditional attitudes and professional insurance provisions inhibit the search for better ways of working.

- NEC procedures can be streamlined once teams are confident in working together in a spirit of mutual trust and co-operation.

- Grouping several projects together into one package makes it easier for contractors and consultants to make the changes required by the Agency's strategy.

- The Agency's strategy applied to a package of projects is delivering improvements in performance equivalent to 20% lower costs and 30% faster completion compared with traditional approaches.

The Environment Agency was established in 1995 to protect and improve the environment throughout England and Wales. It has specific responsibilities for water resources, pollution prevention and control, flood defence, fisheries, conservation, recreation and navigation. An important objective is contributing to sustainable development through the integrated management of air, land and water.

Within its overall annual budget of some £600 million, the Agency is responsible for a £150 million annual programme of flood defence work, of which 12% is spent on consultancy fees, 4% on in-house project management and the remaining 84% on engineering and construction contracts. The Agency was unhappy with traditional approaches to construction, and so right at the start of its work, in 1996, it commissioned Gardiner & Theobald to define a new procurement strategy.

The new approach adopts the same principles as the Egan Report, and specifically aims to achieve continuous improvements in value for money and quality by using partnering wherever possible. Particular effort is being put into improving the planning of projects, establishing more productive relationships and carefully matching the contractual approach to individual project needs. The Agency uses Trusting the Team as its main guide to partnering.

Completed scheme at Crimpsall

Environmental mitigation works to add some
meanders to a previously canalised section
of the River Calder

Wier at Crimpsall

NEC is seen by the Agency as a key tool in putting partnering into effect. A careful review was undertaken of the main approaches to conditions of contract before the decision was made to use NEC. The Agency decided not to produce bespoke contracts as it was thought this would complicate the work of the construction industry. The Agency took into account the fact that NEC is already widely used, is readily available, and future development is supported by a substantial organisation, the Institution of Civil Engineers. They liked its clarity and use of plain English. It was a key advantage that NEC provides the range of procurement options needed by the Agency, so projects can use the most appropriate approach without staff having to learn different forms of contract. It encourages positive supply chain management, which fitted the Agency's aim of reducing construction costs. A key factor was that the NEC is the only form of contract that positively encourages partnering behaviour. As a result of the review, the Agency now uses the NEC for all engineering and construction contracts, including those for professional services, and contractors are expected to use it for subcontracts.

An important benefit emerged in practice, as using a radically new contract forced the Agency's staff to rethink the way they work, and this has made it easier to introduce the overall package of changes required by the new procurement strategy.

Training has been provided for internal staff in the project management skills that the NEC requires. These include benchmarking, cost management, risk assessment and management, and a general understanding of procurement and contract strategy.

The number of consultants and contractors used has been reduced by carefully selecting those who are competent and willing to work co-operatively. The aim is to work with firms who want to develop a long-term interest in the Agency's work and are prepared to extend the Agency's approach throughout their supply chains. One key part of this is joint staff development, whereby new methods are learned in a co-ordinated manner. Particular emphasis is given to cost management, negotiation, co-operative behaviour, team building and proactive project management. This staff development is tied into personal development plans matched to viable career paths that provide real motivation for individuals to buy into the new approach.

Putting appropriate management structures and systems into place took time. To steer the initial implementation of the new strategy, the Agency established a Procurement Initiative Group of senior managers representing all the key internal interests, assisted by a specialist independent advisor. The Group reviewed, improved and then gave authority for new procurement ideas proposed by project managers. A key part of the Group's work was ensuring that project managers took into account all the implications of their specific proposals, e.g. that the use of NEC target contracts was fully understood. The Group also received feedback reports, describing interesting events and the final outcome of all the Agency's projects, to identify lessons for the future. This provides the basis for benchmarks, which are used in setting targets aimed at continuous improvements in performance. As planned, the Procurement Initiative Group was disbanded in 2000 when all the Agency's project teams had been trained in the new procurement strategy.

In addition, regional teams, including representatives of procurement, project management, finance and audit departments, have been established to ensure that local projects are consistent with the new strategy. They have access to external specialists in partnering, team-building, risk and value management, benchmarking, procurement and contract strategy.

A National Capital Programme Manager co-ordinates the work of the regional teams. His job is to drive the application of best practice across all the Agency's projects, and so realise the ambition of being a leading best practice client.

Project managers are a carefully selected mix of in-house staff and consultants. A team approach is encouraged, and selecting the key members of project teams as early as possible helps foster development of this ethos.

Particular attention is given to ensuring that the contractors invited to tender are serious about using partnering. Then the choice of contractor is based on competitive tenders and carefully defined quality and price selection criteria aimed at providing best value for the Agency. It is planned that a framework comprising a limited number of contractors will be doing most of the Agency's work by the end of 2001.

Experience shows that NEC causes project managers to give extra thought to what they are trying to achieve. Completing the Contract Data required by the NEC forces them to think carefully about what they are doing. This is one illustration of the practical effects of the NEC philosophy of carefully planning work before it is carried out. This has become a reality on the great majority of the Agency's projects, with one very clear result being that there are fewer problems to sort out during projects.

NEC causes project managers to concentrate on ensuring that their projects go well, rather than spending time and resources making sure they have a consultant or contractor to blame if things go wrong. Risks are carefully identified and the options for managing them considered before a procurement route is chosen. For all larger projects, the main risks are discussed with the selected contractor before the contract is finally agreed. This often provides surprises for the Agency about how contractors evaluate risks, nearly always gives both parties a better understanding of each other's point of view, and occasionally produces new ideas about managing risks.

The Agency has recognised that partnering does not eliminate the natural human instinct of most individuals to look after their own interests. So, for example, project managers maintain a tough, proactive approach to cost management throughout projects. This is particularly necessary where target or cost reimbursable contracts are used.

**Flood Embankment at
Pateley Bridge**

A 'blame culture' still exists in professional indemnity insurance, restricting what consultants are allowed to do and making them overcautious. The Agency sees a need to find some way of setting up project insurance so that project teams can concentrate on finding the best answers from everyone's point of view.

Partnering and NEC have caused a major cultural change internally for the Agency. The change has been helped by internal auditors who recognise that on occasions it is worth taking a risk to get the best value for the client. The main change is a much more open exchange of information with consultants and contractors in searching for the best answers. In response, contractors now propose better designs and more effective construction methods in a way that never happened when traditional forms of contract were used. A feature of NEC which has proven to be especially powerful in fostering this co-operative behaviour comes from Clause 16.3, which requires those attending early warning meetings to seek solutions that will bring advantage to all those who will be affected.

Current projects have budgets that require cost reductions of between 10% and 20% compared to the Agency's traditional approach. An important example is the North East Combined Capital Project, which is being undertaken on a design and construct basis by Edmund Nuttall with Babtie as design subcontractors.

North East Combined Capital Works Project

The Combined Capital Works Project comprises seven geographically separate flood defence schemes within the Agency's North East Region, grouped into one overall package to provide continuity over a sufficient period of time to justify consultants and contractors investing in the changes that the Agency wants them to adopt.

Based on the Agency's established traditional practice, initial estimates of the costs of the projects ranged from £0.5 million to £5.75 million, with a total for the seven of £11.9 million. The work is to be completed in three years, but the contractor has flexibility in programming individual projects within this overall timeframe, subject only to annual limits on the Agency's expenditure and specific constraints on the timing of some work arising from environmental impact concerns and third party restrictions.

Halcrow was employed by the Environment Agency to advise on the approach that should be adopted. An important factor in Halcrow deciding to advise the Environment Agency to use NEC was that it provides the options needed to deal with the different state of design development at tender stage on the individual schemes.

Halcrow also made sure that the nationally provided training covered all the changes to the engineers' traditional role that the Combined Project required. In addition to NEC procedures, the use of design and construct, partnering and target contracts all needed new skills and attitudes.

Halcrow helped draft the tender documentation, ensuring that risks were carefully considered and taken into account, and helped to evaluate the tenders and interview contractors.

Selection of the contractor

Given the need for the major changes in attitude required by the Agency's new approach, the contractor was selected very carefully. To comply with EU rules, the package of seven schemes was advertised. This identified 49 potential contractors who were sent an initial questionnaire, which encouraged an open dialogue with the Agency. This resulted in six contractors being interviewed and four being selected to tender. A pre-tender conference and site visits took place during the three-month tender period, and all questions raised, together with the answers, were given to all tenderers. Tenders were evaluated using a predetermined scheme that took account of quality issues, including safety, risk management and partnering, as well as cost. The evaluation resulted in two of the four tendering contractors being interviewed before the final selection was made, to ensure that the team chosen had the right co-operative attitude.

Use of NEC

At the time Nuttall were appointed, five schemes had initial designs and so used Option A, Priced Contract with Activity Schedule. The other two, where the Agency's requirements had been established but no design carried out, used Option C, Target Contract with Activity Schedule. The Works Information described the required performance plus the Agency's preferred design approach, for which they had obtained all permissions and approvals. Any detailed design already completed was included for information only and not warranted. Alternative prices were invited for the Agency or the contractor to carry risks, including unforeseen ground conditions and underground services.

Another feature of the project has played a significant part in encouraging Nuttall to look for better answers. This is a provision in the Works Information by which savings that result from changes to the Works Information proposed by the *Contractor* are shared 50:50 between the Agency and *Contractor*. This arrangement has been incorporated as an Option Z additional condition of contract.

Project organisation

The Agency has appointed internal staff as project managers for each individual project, plus a very experienced senior manager for the overall combined project. RKL-Arup has been appointed as the supervisor and is providing some quantity surveying staff, particularly to help deal with the two target projects.

NEC has altered the Agency's administration compared to traditional approaches, e.g. in quality control. Traditional practice would have required three staff to undertake detailed tests compared with the approach used on the Combined Capital Works Project, which requires one supervisor auditing the contractors' own systems. On the other hand, more effort is involved in administering the two small target cost projects than was originally expected. Also, cost control on these projects has required additional quantity surveying staff to keep track of actual costs.

In deciding on their own management approach, Nuttall took into account the wide geographical spread of the individual schemes, and the fact that all except the one large scheme are too small to justify dedicated project teams that include all the required skills. They also took into account that, although the largest scheme has a long construction period, this is due to constraints written into the Works Information rather than the work being technically difficult. Taking all this into account, Nuttall decided to set up scheme teams on each construction site comprising all the people who need to be in day-to-day contact with the direct construction work. In addition, a central project office was established in Leeds, led by a project manager, to deal with design management, planning, finance and overall management of the project.

Babtie established a core team led by a Director to undertake all the design elements, based in their office at Wakefield. Being based in their home office makes it very easy for Babtie to supplement the core team when necessary. For example, one of the schemes includes the design of a dam for which Babtie needed to use the firm's specialist expertise. Also, when the design work needed to be accelerated, Babtie called on extra designers from other teams within the same office. Having an integrated core team allowed Babtie to be represented at all project meetings by people who understood the design of the overall project as well as the individual projects.

As a result of the different way in which each of the main parties needed to organise its own work, it was decided not to establish a common project office. The possibility of basing a common project office in Nuttall's or Babtie's office was considered. At the time, key staff felt it more important to retain close contact with their technical support and key colleagues than to be part of an integrated project team. With the benefit of hindsight these same key staff now feel that a co-located project office would have improved communication, helped them solve problems faster, and made the project even more successful.

In the event, the overall project interest was provided for by setting up a strategic team of senior managers from the three organisations (Environment Agency, Nuttall and Babtie). It meets monthly to review the financial position, progress and public relations issues for the overall package. It deals with problems and generally ensures that the project teams are working to agreed objectives.

Stages in the project

Nuttall and Babtie had developed a programme to the detail required by the ECC as part of their tender submission. When they were appointed in November 1998, their first step was to review the, by now, accepted programme. They considered each individual project to decide exactly how it should be tackled. In doing this they found it necessary to develop more detailed programmes for design and approvals.

Work began on site in January 1999 on the largest scheme. The majority of the design for this was provided by the Agency on an information-only basis. Babtie adopted the existing design and, since the project was not technically complex, it was practical to make an early start on site.

The next scheme tackled were those with defined completion dates. Some design existed for these and effort was concentrated on using it to ensure an early completion. This involved issues affecting local communities, such as temporary road closures, which had to be resolved and approvals obtained.

By starting quickly on the largest scheme and on those with fixed completion dates, Nuttall maximised the chances of satisfying the Agency's time requirements. The remaining schemes were fitted around the early schemes as design was completed, approvals obtained and staff were available.

While work is underway on individual projects, there are separate monthly progress meetings attended by all the key people involved on the particular project. When work is underway on site, the meetings are held on site. These are all held within a few days of each other. The strategic meetings of senior managers for the overall package take place just after the monthly round of individual project meetings.

Design and construct

Design and construct contracts provide important advantages in giving the customer a single integrated organisation to deal with. There is no ambiguity over responsibilities, and so fewer disputes. The integrated approach also often helps construction firms work more efficiently. However, the arrangement needs to be set up carefully to get the full benefits.

The Environment Agency's use of a mixture of full but unwarranted design information on some schemes and only very preliminary design ideas on others has produced its own problems. The difficulties in evaluating the quality and completeness of the provided information in the relatively short tender period inevitably required assumptions to be made that affected design and construction costs. Differences to these assumptions that subsequently developed made the management of the interface between the construction and design difficult.

At tender stage all tenderers expressed concern about costing the less developed schemes and two of these were converted from Option A lump sum to an Option C target cost basis. This action provided more opportunity and incentive for savings to be generated as those less developed design ideas were progressed to construction. It still however, left uncertainties at the tender stage making the setting of the target difficult although this is compensated for by the NEC target mechanism.

The design and construct fundamentally alters traditional relationships between customers, consultants and contractors, which needs to be managed by use of NEC, partnering, workshops and open discussion.

Partnering

Partnering provides a set of actions designed to deal with the type of relationship issues raised by adopting radically new ways of working. The Agency staff in the region understands this in describing partnering as the oil that helps the project management machine work more efficiently. They recognise that partnering means a major change in behaviour on the part of everyone involved. The effort needed to make this change is seen as worthwhile because they recognise that contractors and consults have talents and expertise that can help achieve better value for the Agency's money. In the past, traditional methods may have caused this talent and expertise not to be used for the Agency's best interests. Partnering serves to align everyone's interest in genuine teamwork, but it takes time and effort to make this change. NEC helps in this because its procedures encourage people to adopt partnering behaviour.

The task of building up partnering attitudes began with the initial selection of a contractor and consultant who had experience of partnering and had the right attitude. A two day partnering workshop was held for all people involved in decision making from the Agency, Nuttall, Babtle and RKL Arup as the supervisor. The workshop used an independent facilitator to lead teambuilding exercises, discussions of personal and joint objectives, the production of a partnering charter which records the agreed cost, time, quality and safety objectives, and a problem resolution ladder to ensure that problems do not fester, but are either resolved quickly or referred to a higher level in a set time frame. The charter includes the aim of 'looking for solutions not scapegoats', which reminds people to look for answers that take everyone's interests into account.

The workshop dealt with the principles of partnering, and established high-level objectives that met the aims of senior staff.

The independent facilitator used a well developed structure for the two days that provided an enjoyable shared experience for the team and generated considerable enthusiasm for partnering. It also assisted in raising problems, for example, the workshop highlighted a number of misunderstandings about the responsibility for quality. The Agency had decided to rely on self-certification of quality by the contractor and subcontractors, provided the design subcontractor signed the certificates. Babtle were concerned that the Supervisor's role seemed to duplicate their own responsibility for signing quality control certificates. The workshop helped everyone understand that, given a self-certification approach, the Supervisor's role is to audit the contractor's quality systems and that they are being applied. It was agreed that Babtle's certificate meant that they had done exactly the same and that the Agency wanted the assurance that both the Supervisor and contractor (including the design subcontractor) were satisfied with the quality control. Although this issue was not fully resolved at the workshop, several further meetings recognised that the wording of the certificates needed to be redrafted. Nevertheless, the partnering workshop provided a common understanding of the approach to quality control the Agency wanted and focussed subsequent efforts on finding a practical way of delivering it.

This illustrated what turned out to be the main benefit of the workshop. People got to know each other and understand what they each wanted from the project. This has made it easier for project teams and the strategic team to discuss problems and to find solutions. The initial workshop has been repeated as new people have joined project teams.

Further workshops have helped to provide a better understanding of the different way each of the parties sees certain risks, to consider progress on the largest project in the package, and to identify lessons on completion of the first of the smaller projects. Interestingly, the second workshop included training in NEC procedures by an external expert. One workshop discussed the results of a questionnaire survey of everyone involved in the project, which gave a score for performance in respect of the partnering charter objectives. This produced scores for the objectives, and the workshop concentrated on identifying ways of improving the weakest aspect of the team's work. Joint social and sporting events have been used to reinforce the mutual understanding built up at the workshops.

It is clear that a partnering attitude of making decisions, which take account of each other's interest, has developed between the Agency and Nuttall's staff. Both believe that they understand each other's concerns and interests far better than on previous projects. Important effects of this sensible approach are that discussion are far more open and it is easier to solve problems and find win-win solutions than on more traditional projects. This all encourages trust and results in productive work, rather than wasting time in analysing cost calculations and contract clauses.

Partnering in practice

The good effects were fully demonstrated when a major problem arose with one of the smaller projects where the target cost option was used. The project appeared likely to face a cost over-run of close to 100%. This represented very poor value from the Agency's point of view.

The difficulties had arisen because design assumptions stated in the Works Information at the time the contract was let, were subsequently changed due to model analysis results. The project team was solving the problems caused by this change incrementally, and the cost, in traditional style, was rising relentlessly. The strategic team reviewed the situation and asked the project team to use a value-engineering workshop to rethink the design from scratch. Members of the strategic team encouraging their own staff in the project team to be creative in looking for a better answer reinforced this. The team spent some five to six months looking at many options. The fact that this scheme was part of a package allowed everyone involved to spend this time without creating big contractual problems.

The result was a completely different approach that dealt with the practical difficulties and provided a good answer at 15% above the original budget. It required changes to the Works Information, such as reducing the specified design life of some elements from an unrealistic 50 years to a commercially sensible 30 years. This reduced life-cycle costs and was adopted because it provided better value for the Agency.

The way this cost problem was resolved is unlikely to have been achieved with traditional methods. The scheme was based on a target contract under which a traditional, adversarial contractor could have argued that the cost of the design on which the target was based had doubled, and demanded a share of the large saving produced by the new answer. The team was able to agree that the only fair way forward was to completely re-price the new design and to establish a new target cost. This very practical answer came from the project team having the authority and the willingness to work together to solve problems in ways that take account of all their interests.

The benefits of partnering went further because the overspend has been offset under the strategic team's guidance by savings identified on other schemes. The change in attitude which this all represents has not happened suddenly, but has been carefully built up through workshops and other initiatives designed to help all the teams recognise their mutual interest in achieving successful projects.

An interesting footnote on the benefits of intensive co-operative working is that the scheme, Crimpsall Sluice Replacement, won the ICE Yorkshire Association's 2001 award for excellence in concept, design and construction.

Compensation events

The great majority of compensation events have arisen in response to a notification by the contractor that a problem has arisen and needs to be dealt with. There have been very few compensation events, about 70 across all seven projects when this case study was researched, which was 75% through the currently planned time. The use of design and construct tends to keep the number of compensation events fairly low, as does the transfer of ground condition and underground services risks to the contractor.

The NEC procedures for compensation events are designed to encourage a search for the best answers from everyone's point of view. In general terms, the key people involved have found that the best answer from Nuttall's point of view usually suits the Agency's interests.

However, it was identified at an early stage in the project and raised as a topic for discussion at one of the workshops that compensation events were not being dealt with strictly in accordance with the NEC timetables. It was accepted that the NEC procedures are a positive step forward from traditional methods because they help everyone to resolve financial matters quickly as the project progresses. Nevertheless, the timetables are unrealistic on design and construct projects for compensation events that require design options to be developed and their impact on cost, time and risk evaluated sufficiently to enter into firm commitments without either party taking on undue risk.

All parties accepted that on occasions NEC timetables had not been met and following discussions to identify the problems and possible solutions, measures were put in place to monitor the team's performance, subsequent turnaround times have greatly improved and revised timetables have been agreed where applicable.

Another result of the NEC procedures is that the customer does not always get the best value answer because the speed required does not allow time to evaluate fully all the options.

In many cases, it has been necessary to get on with work before a quotation has been prepared and the project manager has made a decision. This arises from the nature of flood defence work – problems arise and have to be dealt with quickly. The Engineering and Construction Contract provides for this parallel working by using Clause 61.1, but the Professional Services Contract does not have a similar provision.

On balance, although the NEC procedures require a lot of work during projects, both the Agency and Nuttall believe that the benefit of knowing where you stand on costs and time more than compensates for the greater intensity of work.

Packaging

The package of seven projects was put together on the basis of the Agency's budget and environmental priorities. The projects were not selected because they required similar skills or experience. Consequently, the main benefit has been the flexibility allowed to the consultants and contractors in planning seven projects. This flexibility has been used, for example, to provide the time needed to solve difficult design problems before work begins on site. The flexibility has also been used if work is held up on one project, to move workers to another project within the overall package. Individual small contracts simply do not provide these opportunities to make the best use of people and time.

Quality control

The approach to quality control is working well. Nuttall has a well-developed system, which includes a detailed checklist and provision for non-conformity reports. The Supervisor, RKL-Arup, has audited the control system and carries out checks on site to ensure that it is being applied. Very occasionally Arup spots a quality problem, and in all instances Nuttall has dealt with it quickly. As described earlier, Nuttall, Babtie and Arup are all responsible for signing off work when it is complete.

Copies of Nuttall's internal non-conformity reports are sent to the Agency to provide positive assurance that the quality control procedures are being applied.

Babtie audits all the quality control paperwork and signs off the work when it is complete. Babtie is not required to inspect the actual work, but certifies that the contractor and subcontractors have applied appropriate quality systems to their designs.

The Agency felt that the importance now attached to quality was well illustrated by the way Nuttall dealt with a minor leak in a wall on a completed project caused by exceptional floods during the summer of 2000. One of Nuttall's directors commissioned an independent report to determine the cause, which turned out to be a changed design detail, and used the incident to emphasise the absolute importance of everyone taking quality control very seriously at all stages.

Cost performance

Overall, the contract sums for the seven schemes are some 10% lower than initial estimates based on traditional approaches. It is realistically anticipated that the final accounts will be settled within these contract sums, in contrast to traditional contracts, which typically over-run by 20%.

Time performance

The project would have finished nearly 12 months early if the exceptional floods during 2000 and the outbreak of foot and mouth disease early in 2001 had not prevented the final planting work from going ahead. This would have meant a reduction in time of just over 30%, and since these disasters affected all construction projects irrespective of the approach used, 30% probably remains a fair evaluation of the time saving achieved. In part this arises from Nuttall planning from the outset to have sufficient staff to complete four months earlier than the contract completion date. However, Nuttall has been able to improve dramatically on the original plan as a direct result of good relationships established during the project.

The overall time was largely determined by the biggest project within the package, in Hull Docks where the dock owners, Associated British Ports, had insisted that the work be undertaken in small sections, one at a time. ABP's previous experience with contractors led them to expect disruption to their own activities and they wanted to minimise these bad effects. So the requirement for sectional working was written into the

Works Information. Nuttall began the Hull Docks work right at the start of the overall project, planned it very carefully, and made sure the actual work went well. Time was also spent understanding the Agency and landowner's real interests, and keeping them fully in the picture about plans and progress.

All this helped prepare the way for a proposal by Nuttall to be allowed to work on two sections simultaneously to be taken seriously. The earlier completion helped ABP's development plans in Hull, and the faster rate of working fitted the Agency's capital expenditure budgets, so the Works Information was changed. The clause inserted in the Works Information[1] for the Option A contracts, which gives Nuttall a 50% share of savings resulting from ideas they propose that require a change to Works Information, undoubtedly encouraged them to take this initiative.

The Agency and Nuttall are confident that the final account will be agreed within one month of completion because all the big issues are being resolved as they arise. In contrast, the Agency's parallel evaluation of what the effects of using traditional ICE contracts would have been suggests that final account negotiations would most likely have dragged on for months, or even years. The Agency and Nuttall see the way that NEC prevents problems from building up as one of its most important benefits. There is more administration during the contract, but this is more than compensated for by the promise of a quick and clean end to the contract.

Lessons for the Agency

The Agency has learned to make the selection of contractors for future design and construct projects very carefully. In particular, it will question whether a genuinely integrated project team exists, the setting up of a project office for key decision-making will be encouraged and the Agency will check whether sensible procedures are in place for dealing with routine work outside the project office. The Agency will also look for better use of information technology to help communications throughout the whole project team.

An important lesson from the Combined Project is that the works information needs to be prepared very carefully, so that it says exactly what the Agency wants and no more. It is easy to include arbitrary constraints that prevent contractors from adopting the best answers. This is especially the case with design and construct, where the works information should define the end requirements and all the essential constraints without dictating specific solutions. On the other hand, it should include all the customer's requirements. For example, life-cycle costs are important to the Agency, but this was not stated in the works information. The aim should be to give contractors the freedom to find cost-effective answers that fully meet the customer's needs. One idea the Agency has for achieving this is to stand back from the task of drafting works information and think about how it could be seen by a contractor. There is a tendency when people are close to a project for them to make assumptions, without realising they have done so because the matter is entirely obvious to them. This trap could be avoided by asking colleagues to suggest possible designs that meet the performance requirements and check if they would be acceptable.

The North East Combined Project was the first NEC and partnering contract entered into by the Environment Agency. Most staff from all parties were new to the approach used and to the culture it requires, but the outcome in all key areas is very satisfactory from the Agency's point of view. Consequently, the approach is being used for a similar scheme starting in 2001, taking into account the lessons from this case study. Nuttall are again the design and construct contractor, with Arup Associates as design subcontractors.

---

1 Authors' note: The more correct location for such a clause would have been as an additional condition of contract in secondary Option Z.

Picture acknowledgments

Courtesy of the Environment Agency

Costain Limited's experience of NEC and partnering

- Projects that are given sufficient time and resources to prepare detailed works information and evaluate scope of work properly outperform those started quickly with inadequate information.

- NEC is a positive force in encouraging good practice as long as the appropriate option is used.

- Target cost contracts allied to an appropriate allocation of cost savings or excesses are very effective in concentrating project teams on finding the best answers within the customer's budget.

- Joint training for project teams helps in partnering, and specific project procedures help build the team spirit and mutual understanding that is central to successful projects.

- Partnering is best put into effect through properly facilitated workshops.

- Co-operative working is best put into effect by naming the core team as the Project Manager in individual ECC contracts.

- NEC does not require new technical skills, as it makes the most effective use of existing skills.

- NEC forces timely decision- making.

- Experienced teams trusting each other to do work right the first time can streamline NEC procedures.

- A common project team office is very effective in welding people from all the key firms into an integrated project team.

- Teams need to consider when to use information technology and face-to-face meetings in making decisions.

- Performance needs to be measured to ensure continuous improvement.

Costain Limited is part of the Costain Group. This case study relates to their civil engineering projects, which include highways, bridges, tunnels, ports, airports, railways and sewage treatment plants, with an annual turnover of £166 million. Costain seek to become involved in projects early to give customers single point responsibility, integrated planning of all design and construction processes, cost certainty, early completion and value for money. Increasingly, these aims are being achieved through partnering, in both formal partnering agreements and conventional contracts.

Increasing numbers of Costain Limited's customers for civil engineering projects now choose to use NEC. The most enlightened provide training to prepare people for the changes required by NEC procedures, and take active steps to encourage project teams to work in a spirit of mutual trust and co-operation. These customers provide time and resources for project teams to co-operate in looking for the best design solutions, and consider different ways of undertaking the construction. They are rewarded with lower costs and earlier completions and provide contractors with a fair return. All this is in stark contrast to some traditionally run projects, where all too often cost and time over-run. Although contract sums are often little different, administrative costs can be two or three times greater and the final cost can be massively higher.

Many of Costain Limited's civil engineering projects have to be completed within a tight contract period, yet there is limited opportunity to do a full, detailed ground investigation before work starts. As a result, the works information has to be based on general assumptions made by the design consultants. It is only when the actual characteristics of the site and the underground services become clear that the final design can be determined. However, construction work needs to go flat out in accordance with a tight programme to achieve the earliest possible completion. There is often insufficient time to evaluate alternatives to find the best option. It is not uncommon for a decision to be needed within 30 minutes. This inevitably puts designers under pressure to make decisions quickly, and it becomes impractical to comply with the timetables in NEC procedures.

However, the problems can be more fundamental than this, because many of the examples of poor use of NEC result from clients' teams having inadequate resources and too little time to prepare the works information. They nevertheless employ contractors on the basis of contracts that require them to carry a large proportion of the risks. This is usually achieved by using either Option A or Option B (sometimes heavily amended) with works and site information that is neither well thought out nor clearly defined. This provides a poor basis for anyone to use NEC.

On projects set up in this way, the contractor's staff find themselves faced with a stream of changes, consultants with insufficient resources to stick to the timetables in NEC procedures and clients who treat compensation event quotations as traditional claims. Often, revised programmes are immediately out of date and so do not get agreed. Work goes ahead without the contractor's quotation being approved. After the event, there are disputes over whether the best approach was used and, therefore, what are admissible costs. Typically, most of the final account is sorted out after work is finished, in a manner very similar to traditional projects.

A fundamental problem with projects that are set up badly is that rapid and large-scale change invalidates the basis of the contract. It is not possible to evaluate the effect of each individual compensation event because the effects are cumulative. As a result, resources allowed for in the contract price, based on a reasonable interpretation of the works information, turn out to be inappropriate or ineffective as the actual nature of the work becomes clear. This does not result from any one compensation event, but is caused by the cumulative effect of many events. NEC procedures do not deal with these situations, where it gradually becomes apparent that the essential nature of the work is not accurately described in the works information any better than it would be when using other forms of contract.

NEC provides options that deal with each kind of project, but when the wrong option is chosen the outcome is likely to be that everyone involved is forced back into a traditional way of behaving. The inevitable consequences are that resources are wasted, both in unproductive arguments and ineffective ways of working, and so costs escalate and profits disappear.

All of this is in stark contrast to projects that are run well, especially by enlightened clients who use NEC as part of

Duffryn Bridge

a partnering approach. Typically, such clients provide continuity for the consultants and contractors they employ. This is an essential first step in developing real trust and co-operation. As The Seven Pillars of Partnering makes clear, the full benefits of partnering take time to develop and so continuity is crucial. Given the assurance of continuity from a client, Costain feels able to provide continuity for key subcontractors. It becomes worthwhile to have regular meetings with these key subcontractors to learn how they want to work, which one is best able to provide specific facilities, and how to get the lowest costs and best programmes. As part of developing their supply chains in this way, Costain uses the NEC subcontract and encourages subcontractors to train their staff in the new ways of working, often involving them in Costain's own training programmes. It is difficult for contractors to have the confidence or the time to take these initiatives unless their clients and consultants adopt an equally enlightened approach.

In Costain's experience, projects using a partnering approach should begin with joint training for all the key members of the project team. Important parts of this are ensuring that everyone understands NEC procedures and terminology and the specific partnering arrangements and philosophy. However, such joint training also helps build up the kind of team spirit and mutual understanding that is central to all successful projects.

NEC does not require Costain's staff to develop new skills, rather it makes the best use of their existing skills. For example, they always programme work in detail for their own use. NEC simply brings that programme into the heart of the project where it benefits everyone involved. Similarly, they prefer to work in a co-operative manner when this is reciprocated, because they know this provides the best outcome for the client, consultants and themselves.

On well-run projects, the client provides sufficient time and funds to allow consultants to produce proper works information before construction on site begins. The NEC option is chosen carefully, taking account of the nature of the project. On fast-track projects subject to frequent changes, the best choice is Option C, Target Contract with

Activity Schedule. In these cases the target should be carefully defined so that responsibility for specific assumptions about ground conditions, underground services and other risks are agreed and explicitly built into the target.

During the project, the NEC procedures and timetables are adhered to. Problems are discussed as soon as they are identified and the NEC procedures provide a positive pressure to find answers that take account of everyone's interests. Consultants take a sensible approach to compensation events. Where they are minor, they trust Costain to deal with them, taking account of everyone's interests in the best and most efficient manner. This helps create time to deal with major compensation events in the controlled manner envisaged by the formal NEC procedures. Costain's experience shows that when the procedures are used in this way the NEC is the best suite of contracts they know because it provides very positive encouragement for everyone to co-operate in all their joint interests. When this happens, sensible trade-offs can be agreed, the most effective answers can be found, and everyone, including the client, ends up better off.

Accepting these strengths, Costain's staff think NEC procedures could, with advantage, be streamlined. As they stand, they generate too much paperwork and encourage people who prefer formality to efficiency.

Costain's experience suggests that, given continuity, use of the right NEC option, and consultants who are provided with sufficient resources for them to do their work properly, the keys to successful projects are people who know what they are doing, plus lots of discussion and open communication. The results are the mutual trust and co-operation that NEC calls for, prices that provide excellent value for clients and fair profits for construction companies.

Afan waste water treatment works

All the principles of good practice came together on the new £30 million Afan waste water treatment works in South Wales. Welsh Water needed the project quickly and decided to undertake it through a well thought out partnering arrangement between themselves, design consultants, quantity surveyors, main contractor, and key subcontractors and suppliers. The outcome is a very successful project that uses an elegantly straightforward approach.

Quantity surveyors EC Harris produced a notional bill of quantities that provided rates for the type of work most likely to be required. This was used to invite competitive bids from suitable contractors. Tenderers were told that the contract would be the ECC, Option C, Target Contract with Activity Schedule, in which savings or cost over-runs would be shared 50:50 between the customer and contractor. So, in addition to pricing the bill of quantities, which would be used in establishing the target, the tenderers were required to state the fee percentage to be added to their actual cost.[4] Selection was based on the preliminary cost information provided in the bids and interviews of the people who would be running the project. Costain Limited was appointed.

One advantage from the outset was that key people had previously worked together on a smaller project using an earlier and less well-developed version of the methods used at Afan. This previous experience of working together helped the team start quickly. They began with a two-day partnering workshop facilitated by a team from the John Carlisle Partnership, which set up the management structure for the project and agreed a partnering charter. The charter states the agreed vision, values and specific objectives based on the Movement for Innovation's[5] Key Performance Indicators, and values.

4 Authors' note: Readers should note that the fee percentage is tendered for all the main options but it is applied in very different ways. For Option B it is only added to the total Actual Cost of compensation events, whereas for Option C it is applied to all contractor's costs. This could result in a different tendered percentage to that submitted if Option B were used.

5 Authors' note: The Movement for Innovation was established following the Egan Report to lead radical improvement in value for money, profitability and respect for people throughout the UK construction industry. Information is available on www.m4i.org.uk.

At this stage the team had a statement, developed by the customer, of the project requirements in the form of a high-level performance specification. This described the volumes of waste water to be treated and the required standards of treatment, safety requirements, restrictions on outfalls into the river, and similar essential requirements. This included the required completion date, which allowed 30 months to complete the design and construction. The customer's budget was £28 million.

Project organisation

The core team provides the engine room of decision-making for the project. It consists of four people. The manager responsible for the budget and the operations manager who will run the completed treatment works represent the customer. The other two members are the design consultant's design manager and Costain's project manager. In addition, the quantity surveyor's commercial auditor was invited to attend core team meetings.

To emphasise the central importance of teamworking in partnering, it was decided to name the core team as the *Project Manager* in all the consultants', contractors' and subcontractors' NEC contracts.

Within the core team, the core team leader is Costain's project manager, and the customer's representative is the manager responsible for budget. Having the operations manager involved in the core team from the outset helped ensure that design decisions took full account of the end user's needs. Indeed, the composition of the core team ensured there was a strong, consistent focus on the customer's interests.

There is also a strategic team, chaired by the quantity surveyor, which consists of the immediate bosses of the core team members. It meets every three months, and more often if required, to take a general overview of the project, review the current earned value analysis and make any necessary strategic decisions. The strategic team is named as the *Adjudicator* in the main contracts arising from the project, but has never been called on to act as *Adjudicator* and generally has been confident in leaving the core team to run the project. The named *Adjudicator* in partnering subcontracts is an appropriate subgroup of the strategic team plus a representative of the subcontractor. Here again it was never necessary to use this provision as teams at all levels concentrated on anticipating problems and finding robust answers before they could give rise to disputes.

Design, target cost and programme

As soon as the core team was in place, a common project office was established for the customer, design consultants and contractor. It was based in Welsh Water's offices in Swansea, which are in the same building as the design consultant's offices. Key subcontractors and suppliers were brought in as their work became central to the project. The first task was to carry out the initial design work quickly and to undertake early pre-target value engineering studies that identified savings of £2 million. This produced a scheme that met all the customer's requirements, including the firm budget of £28 million. Once scheme approval had been achieved, design and target cost compilation progressed.

The statement of customer's requirements and an initial budget based on the tendered cost data were the starting point. Design was a team effort, concentrating on critical decisions to a point where a preliminary design describing the overall scheme and a target cost of £28 million were agreed. In parallel, an overall programme was developed aimed at meeting the customer's required completion date.

The target cost was based on measured quantities produced by the quantity surveyors from the designers' preliminary drawings,

Steering team

The steering team establishes the overall project strategy, advises on the implementation of policies and acts as adjudicator by:

1 setting policy,
2 promoting partnering,
3 encouraging feedback,
4 maintaining an overview of the core team,
5 ensuring necessary resources are available,
6 resolving issues referred by the core team, and
7 adjudicating on disputes.

Core team

The core team manages the project and implements and encourages partnering in accordance with the Charter by:

1 managing the project,
2 encouraging, implementing and measuring partnering success,
3 authorising expenditure,
4 encouraging and supporting opportunities for improvement,
5 making technical and procurement decisions,
6 exercising financial and programme control,
7 early identification of barriers,
8 encouraging relationships based on mutual trust and co-operation,
9 co-ordinating the project with the end user,
10 managing public relations,
11 publicising the project approach, and
12 resolving disputes.

plus the results of a team review of each drawing to identify work not shown. The quantities were priced at tendered rates or rates negotiated on the basis of tendered rates. About two-thirds of the target was based on tendered rates and one-third on negotiated rates.

Throughout this stage, risks identified by anyone were listed in a risk register. This provided a list of just over 50 risks, including ground conditions, elements of the design still to be defined, weather, failures of design or processes, design development, planning and other approvals, environmental issues and testing. Just before the target cost was agreed, the risks were valued and then reviewed by the core team to determine the size of the risk and the probability of it occurring. These two factors were used to rank the risks, and the core team then decided how to deal with each one. The first decision, which was primarily customer led, was whether the risk should or should not be included in the target cost. In general, risks with a low probability of occurring but high costs if they did occur were not included. The effect of this was that if they did not occur no one incurred any costs. However, in the unlikely event that such a risk did occur the actual costs as defined in the NEC were reimbursed by the customer.

The second decision applied only to risks included in the target cost. It began with the team assessing mitigation strategies for each risk and in doing this re-evaluating the size of the risk and the probability of it occurring. This led to three categories of included risks: an allowance for a defined risk, an allowance for a mitigated risk, and a provisional sum. The first two categories were dealt with using normal NEC provisions. The provisional sums related to those risks where insufficient information was available to establish a price. They were dealt with as compensation events, ideally by agreeing a target cost when sufficient information became available. However, for those risks where this was not possible before the work was carried out, the actual cost was set against the provisional sum in the target cost. As a result, there was no effect on the shares of cost savings or excesses for these risks.

The outcomes of this stage were a target cost of £28 million based on the customer's requirements, an outline design and an agreed risk analysis. The target cost was linked to a comprehensive outline programme for design, construction and commissioning.

Contracts

These outcomes formed the basis of the main contract, which was based on ECC, Option C, Target Contract with Activity Schedule. In the spirit of partnering, delay damages were kept to a minimum and there was no retention. Savings and cost over-runs were shared 50:50 between the customer and contractor. The detailed performance statement of the customer's requirements provided the works information. This had the beneficial effect that there were few compensation events. Apart from the provisional sums dealing with risks, only nine compensation events were needed to deal with changes to the customer's requirements. There was of course a great deal of design development which changed the actual costs, but since this did not change the customer's requirements the target was not changed, and so design development did not give rise to compensation events.

The design consultant was employed on a PSC, Option A, Priced Contract with Activity Schedule. Additional conditions were used to provide for the design consultant to receive a small part of the customer's share of savings or cost over-runs. This arrangement was successful in aligning interests, fostering co-operative working and concentrating efforts on looking for cost effective answers.

Subcontractors' contracts use the ECS, mainly Option B, Priced Contract with Bill of Quantities. A few use Option A, Priced Contract with Activity Schedule. The key contracts included additional conditions to provide for the subcontractor to share 50:50 in savings resulting from ideas they suggest. The most critical of the subcontract work, the control systems that are at the hub of the mechanical and electrical processes, uses Option C, Target Contract with Activity Schedule. This work is critical because nearly every change to other elements has some influence on the control systems, and so the subcontractor was given the same broad incentives as the main contractor to search for the most cost-effective answers.

Welsh Water's Afan waste water treatment works in South Wales

Design and construction on site

Once the target was agreed and the main contract in place, the common project office moved to site, where construction began quickly. An important principle in staffing the project office was to create a single team that included all the expert knowledge needed to undertake the direct work, but which avoided duplication so that everyone on site had a real contribution to make to the project. Establishing a single integrated project information base reinforced this. The sense of team spirit is expressed in the sign outside the project office, which reads 'Afan Partnering Team', and displays the team logo, which is also included on individuals' business cards.

The common project office has made a major contribution to the project's success. Working closely together helped everyone understand current progress and all the factors influencing the work. This meant that decisions were well informed and made quickly, and it rapidly became clear to everyone involved that prompt decision-making was saving substantial amounts of time and resources. It allowed the project to be undertaken rapidly with design work staying sufficiently ahead of construction to allow NEC procedures to be used to maintain control.

The bulk of the design work was done in the designer's Swansea office, where the common project office was originally based. However, two designers were based full time on site. They helped interpret design information and liaise with head office designers to ensure that well-informed decisions were made quickly. The whole contractor's team from project manager down were based on site, as were key staff from all the subcontractors currently active on the project. Staff who will operate the completed plant were also based in the common project office for about two-thirds of the construction stage. This helped ensure an unusually smooth handover, because the operations staff knew the plant in detail and understood how it is intended to be run. Costain's key process staff staying on site for the important first three months of operation further helped the handover.

Time

The NEC procedures provided an effective basis for the very detailed programme that was used to ensure that time was taken seriously by the customer, designers, contractors and subcontractors. The NEC-based programme covered all stages of design, construction and commissioning, and was a major factor in ensuring that this fast-track project stayed under control. Indeed, it did better than this in demonstrating considerable flexibility when the customer introduced a major change fairly late in the project.

The late change was a decision to use the plant to treat sludge imported from elsewhere. This required the sludge drier to be increased from a two to seven tonnes per hour water evaporation drier, and access to be provided for imported sludge. This meant an extra £1.7 million of work. By the stage the change was decided on, the programme was well established and the additional work required substantial rethinking and reprogramming. After much intensive planning, and flexibility on the part of the subcontractors most involved, the compensation event was agreed with no delay to the date for starting operation of the plant.

Cost

The project adopted a very straightforward approach to costs that is broadly based on the NEC philosophy. The target cost established just before the main contract was agreed reflected the customer's requirements. It was changed if the customer changed these requirements, using the ECC compensation event procedures. The notification of compensation events procedures worked well. More importantly, costs were agreed, not at defined Actual Cost as the ECC requires, but on the same basis as the original target using Costain's tendered or target rates wherever possible.

The contractor was paid actual costs, based on his normal accounting system, plus the fee percentage established by Costain's tender. The accounting system was run from the common project office on site and directly recorded all labour, materials and plant used. These actual costs were coded to the main elements and subelements used in building up the target cost to make cost control straightforward. The accounts were open to the core team and subject to monthly audit by the quantity surveyors to make sure there were no hidden cost problems. Also, Welsh Water's internal auditors have examined the accounts on two occasions. None of the audits picked up any significant error in the accounts.

The target mechanism has allowed the core team to concentrate on looking for the most cost-effective answers. The focus on proactive cost control was very necessary on this project, where firm design decisions were made close to construction being undertaken. The ECC target contracts require the contractor to provide regular reports of the forecast actual costs. Cost control was based on this provision to ensure the core team always knew the cost consequences of designs before they were accepted. At the same time, an earned value analysis related actual costs to the target costs to ensure that the core team concentrated on directly productive work.

The simple direct approach to cost control and management used at Afan was a big factor in ensuring that no time was wasted on unproductive disputes over money.

Quality

The contractor was named as the *Supervisor*, which neatly reflects best practice quality management, where the person doing the work is responsible for quality and there is no external checking Costain's quality control system had been seen to work well on the previous project for the same customer with the same design consultant. It was decided, therefore, that there would be no benefit in paying for additional, independent checks.

The customer and designers make occasional spot checks on the quality control paperwork, but there is no formal inspection by them of the construction work. In practice, if site-based designers saw a defect or were concerned about any quality issue, they told the project manager informally. This is consistent with the reality that the whole team is responsible for quality.

The approach to quality control worked outstandingly well. Defects were put right as soon as they were spotted, so there were very few defects at handover.

Partnering

By the end of the project, the team had learned that partnering means open communication, being honest, looking for answers that meet everyone's interests, and being willing to make compromises for the common good. Informal, open communications across disciplines and companies is the key difference from traditional approaches.

In putting this into effect, the team uses a sophisticated mix of information technology and face-to-face communications. It has become apparent that projects as fast as this one require managers to make careful decisions about which form of communication to use for which activities. In general, information technology is used to brief people on current progress or problems. This can lead to questions and further exchanges of information using information technology. However, decisions are made face to face in informal discussions, meetings or workshops.

The initial workshop for the core team established the basis for partnering to flourish. The project's unconventional use of NEC was designed to put partnering into effect. A quite central part of this was the formal project manager's role being undertaken by the core team, which meant that the members of this team had to co-operate on a daily basis. They also held regular weekly meetings to discuss and resolve issues not dealt with on a day-to-day basis. These covered matters such as programme, money, safety and third-party interests. The approach which emerged was that, after discussion by the core team, the person best able to tackle each specific issue did so with the authority of the rest of the team.

The required behaviour and attitudes were further reinforced by stating in tender invitations to subcontractors that the project used partnering. This was explained in detail at pre-tender meetings with each tenderer to clarify the precise role and responsibilities they would take on if successful. The meetings were completely open, so the subcontractor had every opportunity to know exactly what they had to do in what circumstances and conditions before they entered into a contract. Once appointed, key subcontractors' staff attended a one-day workshop designed to allow them to experience the project's partnering approach. The core team often needed to overcome defensive attitudes, as most subcontractors suspected that the workshop was designed to reduce their price. Once this legacy of traditional approaches was overcome, the vast majority of subcontractors fully engaged in the new way of working. As work got underway there were regular meetings to review progress and solve problems. The guiding principle in working with subcontractors was that good communication is the key to partnering.

Motivation was reinforced by the common project office, which made a big contribution to partnering. It showed that when competent, well-motivated people work closely with each other every day on a challenging project, real team spirit develops.

The John Carlisle Partnership has periodically provided a health check on the project's partnering performance by means of questionnaires and interviews at all levels on and off site. The health checks cover attitudes and behaviour, information systems, communication, design and construction processes, and performance benchmarks. They show that partnering progressed from being rated as good initially to becoming very good. In John Carlisle Partnership's very experienced view, this is one of the best projects the firm has ever seen.

Conclusions

The project was handed over to the customer on time and within budget. This performance needs to be judged against the initial time and cost targets, which were widely considered to be tough and would not have been achieved by traditional approaches.

The project is widely seen as an outstanding success and the NEC-based partnering approach is being repeated on subsequent projects. Also, the lessons are being made available by treating the Afan waste water treatment works as one of the Movement for Innovation's demonstration projects.

Costain sees this mature use of NEC and partnering as the right approach for one-off, fast-track projects. It helps them meet customer's needs, work co-operatively with consultants and subcontractors, and earn fair profits. It also helps everyone involved to be happier and less stressed at work. It also, as this project demonstrates, provides substantial improvements in productivity in the engineering and construction industry.

Picture acknowledgments

Courtesy of Costain Ltd

The Halcrow Group's approach to NEC and partnering

- Partnering with customers and contractors is central to Halcrow's business.

- NEC target contracts are the best available form of contract for partnering.

- Experience shows that concentrating on quality and time is the most effective way of ensuring cost efficiency.

- Quality control is best based on self-certification rather than masses of independent checking, where the contractor fully 'buys into' the idea.

- The Channel Tunnel Rail Link, the first major new railway in the UK for over a century and designed and managed by Rail Link Engineering (a consortium of Arup, Bechtel, Halcrow and Systra), is a massive £5.2 billion project using NEC and partnering as central features of programme management and budget control.

- Consultants are paid in a similar manner to contractors, based on the performance of the whole project, which gives them a direct interest in the contractors' success.

- Time and resources spent completing the design before construction begins allows the best design answers to be found and saves massive amounts of wasteful paperwork dealing with compensation events and revising programmes.

- Contractors for individual sections of the work are selected for technical, commercial and partnering skills, and employed using NEC target contracts.

- Common project team offices are used for the overall project and for each main section and they have been effective in building co-operative team spirit.

- The *Project Manager* for individual contracts is also the *Supervisor* to ensure there is no gap between responsibility for design and supervision.

- Contractors are paid right up-to-date at the time the payment is made so they are not required to finance the project.

- Partnering attitudes are established and peoples' expectations raised by facilitated workshops and encouraged by social events and a regular project-wide newsletter.

- Partnering needs constant attention to ensure that everyone co-operates in the best interests of the whole project.

- Meetings bring the key people together with all the information they need to make timely decisions.

- Senior managers regularly 'walk the job' to be visible and available to the workforce and get first-hand knowledge of progress.

- Agreed project procedures are written down and kept up to date as better ways of working are found.

- Consultants and contractors have been able to reduce staff in the common project team office for one section by 25% by eliminating unproductive administration and making the project team a single legal entity responsible for the section.

Track laying at Fawkham Junction

The Halcrow Group is a major international engineering consultant that believes partnering is central to the development of its business. As evidence of this, there is a Partnering Champion on the main board, providing a focal point for training and focus groups, which ensure the spread of knowledge of partnering throughout the company.

Halcrow has extensive experience of NEC and believes that it is by far the best form of contract to use when the customer and contractor want to have a partnering relationship. In Halcrow's experience, NEC operates satisfactorily without partnering, but there are further benefits when partnering is used. Partnering has to overcome traditional commercial attitudes, which view problems for the other party as opportunities for claims. NEC breaks out of this traditional adversarial behaviour by the simple device of stating that the parties will undertake specific actions by defined times. As a result, each knows how the other will behave if they stick to what they have agreed. The contract provides remedies for when either party fails to act as required by the contract, and the resulting certainty encourages co-operative behaviour and trusting attitudes.

Halcrow's experience of NEC suggests that in most circumstances, target forms of contract work best of all. Fixed price contracts seem to leave the door open for traditional adversarial attitudes to reappear. Both NEC and partnering depend on the right incentives being in place. Neither work when the financial arrangements create conflicting objectives. Risk management is important in establishing targets that ensure risks are shared, so that everyone has an incentive to minimise the cost and time effects.

An important issue in managing designers' work is to define efficiency in terms of the overall effects. It is common in traditional practice for designers to think of efficiency solely in terms of reducing the quantity of material used – it is often more efficient to use more material in a simpler design that is easier to construct. In designing incentives for designers, it needs to be kept in mind that targets should relate to overall project costs, not just to the designers' costs.

NEC encourages proactive management by forcing people to make early decisions when they become aware of problems. This is reinforced by the detailed procedures. For example, NEC's requirement for early warnings means that problems have to be identified early, thought through and decisions made. This is closely allied to risk management – problems are identified early and then taken into account in making plans. ICE contracts do not do this. ICE contracts provide effective forensic tools for discovering who was wrong, but give no help in deciding what actions need to be taken to solve today's problem.

In using the NEC compensation event procedures, Halcrow uses the contract timetables flexibly. In general, the timetables create pressure to deal with changes quickly, which is beneficial, and most events can be dealt with well within the maximum times allowed. However, a few major events need more time to allow alternative design solutions to be explored. In these cases, an extended timetable is agreed at the time the compensation event notification is issued.

Putting NEC into practice requires constant attention, because when problems arise many people fall back on their old adversarial ways of working. This happens in the public sector because people are concerned about the view of auditors if a risk is taken to find the best answer. People are reluctant to suggest ways in which they can help contractors be more efficient. There are also problems arising from public sector annual budgets, which limit attempts to improve value for money and efficiency by requiring work to be let as series of small projects. There is now more flexibility arising from block grants, but old habits are slow to change. Also, there is too much reliance on tying contractors to competitively determined fixed prices, even where work has to be designed during the project by independent consultants or in-house public sector staff.

Private sector organisations can also be slow to change. Customers often underestimate the demands of project management, and contractors can be complacent about quality assurance systems because traditionally they have relied on the Clerk of Works.

Achieving quality

Halcrow gives a lot of thought to how quality control systems are set up. Experience has taught them that independent inspection on its own is not effective. It is essential that the people undertaking work feel responsible for its quality. Therefore, as long as contractors have good quality control systems in place, Halcrow relies on self-certification by contractors, backed up by an audit of their quality-control systems and NEC's disallowed cost mechanism.

Quality, however, has to begin with design consultants' own quality assurance systems. They not only define what is required for design quality, but also provide an essential framework for defining what contractors quality assurance systems should provide, and what consultants need to witness prior to and during construction on site.

Halcrow emphasises the importance of quality by putting the quality assurance requirements at the front of the works information. In determining the requirements, Halcrow undertakes a careful risk analysis that takes account of health and safety issues during construction, as well as the design of the end product. In general, Halcrow requires contractors to have a quality plan, general quality control procedures, and specific procedures for each element and system. Halcrow reviews the contractor's procedures sufficiently to ensure that they include everything they need the contractor to do. This is followed by two physical audits. First, to check that the contractor is doing what the procedures say will be done. This is carried out partly independently and partly jointly with the contractor. Second, Halcrow ensures that the procedures are being effective by making random checks of the finished work, again partly independently and partly with the contractor.

When some aspect of quality is wrong, the first question is whether this raises a safety issue. If it does, the work may well need to be stopped, but generally Halcrow tries to ensure that work on site is not interrupted. Nevertheless, every quality problem is investigated to identify whether the cause is deficiencies in the procedures or in their application. This frequently leads to further training for the contractor's staff in quality procedures. Key principles in this are ensuring that contractor's staff make a plan for inspection that varies to ensure that the same things are not looked at in the same sequence every day. Then, when a problem is identified, following it right through until a robust solution is found that makes it very unlikely that the problem will recur.

Halcrow knows that the way the first quality problem is dealt with will set the tone for quality control for the rest of the project. So the maximum effort is made at the start of projects to ensure that everyone takes quality seriously.

People are not allowed to work on site until they have attended a health and safety course, which includes quality control procedures. Sites commonly issue hat badges on completion of the course to make it easier to check that there are no untrained people on site. All this is reinforced on site by holding regular competitions between teams on specific aspects of quality or safety performance. The cash prizes are usually given to charity, but teams like winning. Site newsletters routinely celebrate the winners and emphasise quality and safety issues.

Where the contract payment mechanism is based on completion of activities, quality can be tied in by defining completion as inclusive of completing the relevant quality procedures. This means that if one of Halcrow's spot checks shows that all the paperwork needed for quality control is not in place when a certificate is issued, the contractor risks not being paid. This is further reinforced by the provisions that allow costs to be disallowed where the contractor does not give an early warning of a problem.

In putting this self-certification system into practice, Halcrow has learnt that it is important to involve the contractor in design and to involve the designer in construction. Both designers and contractors have specialist knowledge needed to produce practical design information in the right sequence, and to ensure it is correctly interpreted on site. So, for example, on design build contracts they insist that designers sign construction quality certificates. They do this because designers have to make assumptions about ground conditions, workmanship, tolerances and other matters, and it is important for them to be able to see whether contractors understand their assumptions and if they apply to the actual situation on site.

This approach ensures that designers know they must continue to be involved throughout the construction stages, and that they have the authority to insist on proper quality standards. Many lessons for designers and contractors emerge from this joint working, which is why continuity in partnering (which is commonly called 'strategic partnering') provides a robust basis for improved performance. Halcrow's view is that quality depends on long-term planning and certainty of workload. This means that customers, especially central governments, have a key role in creating the right circumstances for contractors and designers to invest in quality.

Emphasising quality does not mean that controlling time and cost are ignored. Indeed, Halcrow takes the view that quality, time and cost all have to be under control because if, for example, cost over-runs or work is running late, a common response is to skimp on quality. Halcrow's views on these matters are similar to those found in leading Japanese practice, where effort is concentrated on carefully planning all aspects of work and then ensuring that quality is excellent and programmes are put into effect rigorously. The common result when well thought out plans are systematically put into effect is that cost also works out as planned. This is in stark contrast to traditional British approaches which too often focus on cutting cost to the detriment of quality and time performance. There has to be balance because any overemphasis on quality, time or cost usually leads to failure in all three.

An important example that demonstrates these relationships and uses NEC and partnering is provided by the design and construction of the Channel Tunnel Rail Link.

---

6 Authors' note: Another method of ensuring integration of design into construction is to appoint the designer as the Supervisor. See the ABSA Bank case study in Chapter 3.

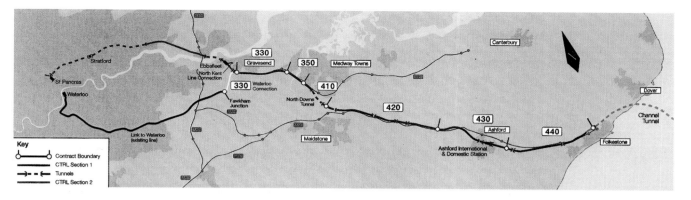

Route of the Channel Tunnel Rail Link

The Channel Tunnel Rail Link

The Channel Tunnel Rail Link is a massive engineering project, providing the UK's first major new railway for over a century. London & Continental Railways are responsible for the design, construction, operation and finance of the project.

The work is being done in two sections, for which London & Continental Railways have established two subsidiaries: Union Railways (South) for Section 1 and Union Railways (North) for Section 2. The outturn costs for Sections 1 and 2 are £1.9 and £3.3 billion respectively. Railtrack will be the future owners of Section 1 and operators of the entire CTRL.

Section 1 provides 74 km of high-speed rail link from the Channel Tunnel to Fawkham Junction in Kent, where it connects to the existing Railtrack network. This includes connections into the existing Ashford International Station. Construction began in October 1998 and is planned to be complete in 2003. It will reduce journey times from Waterloo to the Tunnel by 20 minutes, resulting in a Waterloo to Paris journey time of 2 hours 35 minutes.

Section 2 main works begin construction in mid-2001 to be complete by the end of 2006. It is different in nature to Section 1, taking the new line a further 39 km, much of it in tunnels under the Thames and through east London to St Pancras. Once complete, it will halve current journey times from St Pancras to the Tunnel to 35 minutes making the journey time from St Pancras to Paris 2 hours 20 minutes.

The project includes 26 km of tunnels, 2.5 km of viaduct, 41 rail bridges, 53 road bridges, 23 footbridges and two new railway stations. It will require 12 million m³ of excavation and a similar quantity of fill. Care has been taken to minimise adverse environmental impacts.

Designers and project managers

Included in the London & Continental Railways competitive proposal for the design, project management and commissioning of the Channel Tunnel Rail Link was the appointment of Rail Link Engineering, a consortium of four major companies – Halcrow, Arup, Bechtel and Systra as designers and project managers. These companies are also shareholders in London & Continental Railways.

The 2nd edition PSC was not available at the time the consultant's contract was set up, so a traditional form has been used. However, the contract provides for the consultant to be paid all actual costs plus profit and a bonus (or penalty) paid against a target outturn cost for design and construction. This arrangement gives the consultant a direct interest in the contractors' success.

Construction contracts

The project managers, Rail Link Engineering, reviewed all the available forms of construction contract and in consultation with Union Railways decided to use NEC, Option C, Target Contract with Activity Schedule, for all the main construction and systems contracts. In addition, there are many smaller contracts for items such as ground investigation, ecological work and the advance work needed to allow the main contracts to begin efficiently. These use the ECSC wherever possible.

The NEC system was chosen because its procedures make co-operation a reality. A lump sum contract would have been a nightmare given the early state of design information caused by well-publicised funding difficulties at the start of the project. NEC target contracts provide the right mix of incentives, flexibility and control.

Tying the consultant's bonus to the contractors' performance means that everyone involved has an interest in effective project management. For example, it is in everyone's best interests to avoid problems on one contract having detrimental knock-on effects on the others. Therefore, it may be necessary for the *Project Manager* to instruct a contractor to accelerate some work to make up for problems on other contracts. To ensure that this option is always available to the *Project Manager*, the part of the NEC, Clause 36.2 that allows the contractor to refuse to accelerate work is not included in the Channel Tunnel Rail Link construction contracts.

In NEC terms, Union Railways is the *Employer*, and Rail Link Engineering the *Project Manager* and *Supervisor*, but to give effect to this the term '*Supervisor*' is not used in the contracts. The decision to link the project management and supervisor roles was much influenced by the investigation into the Heathrow Express tunnel collapse, which criticised the separation of responsibility for design and supervision.[6] On the Heathrow project, individual responsibility had become difficult to determine, especially where permanent and temporary works interacted. On the Channel Tunnel Rail Link, the scale of the work makes it inevitable that separate people undertake design, project management and quality control, but there are no contractual barriers between them.

The other major change to the NEC construction contracts is that the first clause (Clause 10.1), which requires the parties to act in a spirit of mutual trust and co-operation, has been deleted at the insistence of lawyers working for funding institutions who considered it unenforceable. Instead, it has been placed in the instructions to tenderers which tell them that partnering is being used. There is no mention of partnering in the formal contracts, other than in the recitals of the Form of Agreement where a statement regarding mutual trust and co-operation appears.

Rail Link Engineering evaluates competitive tenders on a consistent basis, which takes account of technical, commercial and partnering factors in the selection of contractors. These are weighted to take account of the specific needs of each project. A key part of the evaluation is a presentation by the contractor's team, who will undertake the work if they are successful, followed by a question and answer session. Selection is based on the most economically advantageous bid, which is often not the lowest bid.

The partnering criteria take account of the contractor's partnering experience, their key staff's approach to partnering, track record in innovation, flexibility, style of communication, and the extent to which a real team spirit exists within the contractor's team. This last point is especially important in a joint venture bid where the separate firms have little experience of working together. Questions about common systems and methods, share of risks and rewards, and similar issues are especially important.

Works information

Rail Link Engineering decided that, ideally, the works information should be at least 60% firm before tenders were invited, and 80% firm before contracts are signed. On Section 1, this has not been achieved. Contracts are based on full, but not final, information. Many of the assumptions in the contract documents have to be changed as firm design information is issued. As a result, a distinctive feature of the project is that a great deal of time and effort is needed to deal with the many compensation events that have arisen.

Partnering

An early decision was taken to use partnering to help establish integrated teams, which aim to ensure that no surprises are created for the customer. It was expected that partnering would encourage contractors to suggest ways of reducing costs and times. Some such innovations have emerged, but there has not always been sufficient time between initial design and construction to explore many alternative ideas.

Partnering began at workshops attended by all the key stakeholders at each main stage. These used an independent facilitator to help each team establish co-operative ways of working. Initially, the independent facilitator, John Carlisle Partnership, worked with the key people in the customer's organisation and the consultants' joint venture. This concentrated on team building, understanding each other's interests and

Cut and cover construction of tunnel portal

establishing mutual objectives. It produced a mission statement and a partnering charter.

Subsequently, the key people involved in each separate construction contract have held their own start up workshops with the help of an independent facilitator. As well as formal workshop sessions, mountaineering and sailing weekends have been used to build team spirit. The construction contracts each have their own partnering charters, which set different aims for the separate teams. Although Rail Link Engineering encourages partnering it does not seek to impose specific partnering objectives. Therefore, individual charters reflect project teams' attitudes and experience as well as the nature of their work.

Other initiatives designed to foster partnering are described in a project-wide newsletter published quarterly. The Summer 2000 edition describes a major milestone achieved ahead of programme, emphasises the importance of safety and quality, describes a team-building weekend, and has features on project sports events, environmental achievements and links with schools. It is professionally produced and helps build a well-founded pride in the project.

A good level of co-operation and trust is evident throughout the project, but this varies from contract to contract, mainly on the basis of individual attitudes. Where key people behave proactively, partnering is a reality; but when they are defensive about their own interests, work is made more difficult and the results fall short of what could be achieved.

Project organisation

A common project team office for the customer and the design and project management consultant has been set up in Central London. The common project team office also accommodates key staff from contractors involved in design work. Typically it accommodates about 800 people, 100 from the customer and 700 from Rail Link Engineering and key contractors. Staff from all the firms use common methods and procedures.

The common project team office undertakes design and project management through to the start of construction work. Then, as each major construction contract gets underway, a common site office is set up. As construction on Section 1 got into full swing, the common project team office for construction activities was relocated to be on site in Kent. The original common project team office currently houses the team working on Section 2 of the overall project.

Integration in the common site offices varies. In some contract teams the separate disciplines stay apart, but the best have achieved a very integrated way of working. The Ashford contract, which is described in more detail later in this case study, provides a good example.

Project meetings

The customer and senior project managers meet with the senior site managers from each contract monthly, to review progress and deal with problems. Big compensation events tend to dominate these meetings.

Senior managers from all the main companies involved meet regularly to review progress and tackle difficult outstanding problems on all the contracts. A central part of these meetings is an up-to-date earned value analysis for the whole project.

Payment of actual cost

Payment is based on the NEC approach to defined Actual Cost plus Fee, but the Channel Tunnel Rail Link construction contracts are more generous than the standard form. The contracts have been amended to provide for payments based on predicted Actual Cost seven weeks ahead. This means that if the prediction is correct, payments are right up to date when the contractors receive them, because seven weeks is the time allowed for certificates to be issued and payments made. So the contracts are fully financed by Union Railways.

Opening a separate bank account for each contract, which is audited by Rail Link Engineering, has helped in accounting for the Actual Cost for payment purposes. This provides clear open book accounting, which has been an important factor in building up trust.

There were initially some difficulties over the Actual Cost of the contractor's Equipment. It has been decided, for contractor-owned Equipment, to pay the purchase price less the resale price on completion of the Equipment's work. For hired Equipment, Rail Link Engineering accepts the hire rate actually paid.

In general, Rail Link Engineering takes the view that target contracts provide an adequate incentive for contractors to be reasonable about costs, and so having understood and accepted their accounting systems, they consider it safe to rely on spot check audits of the contract bank accounts.

Compensation events

NEC's early warning provisions have worked well, but those dealing with compensation events much less well. All the contracts have got off to a bad start in dealing with compensation events. This is because at the time contracts were awarded, design information was being produced at a rapid rate and contractors tended not to have sufficient staff available to make a fast start. As a result, a backlog of typically 20 or 30 compensation events built up before the team started to work effectively.

NEC timetables are very tight on work of the Channel Tunnel Rail Link scale. It is often difficult to decide quickly if a particular event will have sufficiently large effects to make it worthwhile spending a lot of time evaluating the compensation event. Consequently, contractors are forced to treat every change as a potentially serious compensation event to protect their own interests if the consequences do turn out to be significant. The pressure to generate large numbers of compensation events (and there have been hundreds) is exacerbated by contractors producing very detailed programmes. These typically show about 3000 activities arranged in levels to provide detailed control for individual work elements, subelements and subsubelements. The construction contracts ask for programmes to be updated every four weeks, and it has proved extremely difficult to maintain such detailed overall programmes up to date in a rapidly changing situation. Tying this into the compensation event process has in no way helped.

Whatever the causes, a massive amount of time and resources are being devoted to dealing with compensation events. Senior management often makes the point that this only changes the target, not the payment of Actual Costs. It does of course change the contractor's share of savings or cost over-runs, and so money is at stake. Nevertheless, Rail Link Engineering recognises that a way must be found to facilitate the administration of compensation events in order to better manage the cost of change and so that time could more usefully be spent looking for better designs and cost savings.

One very positive feature of the way compensation events have been dealt with is that the overall completion date has been maintained. When a contractor's quotation shows an increase in time, this is negotiated back to the original timetable. It has not been necessary to use the acceleration clause in the contracts in doing this. This approach reflects a firm conviction on the part of Rail Link Engineering that controlling time is a crucial factor in keeping costs down. They know that once one part of a project is allowed to run late, the knock-on effects can rapidly undermine the whole programme.

On Section 2, it has been decided to let contracts six months before construction needs to begin to concentrate resources on bringing the design to a high degree of completion (the factor that led to a large number of compensation events on section 1) and identify cost savings in the design stage by encouraging designers and contractors to work together in value engineering exercises so that robust design information is completed well ahead of construction.

Rail Link Engineering sees the ECC provision, by which changes to the works information proposed by contractors do not become compensation events, as an extremely useful incentive for contractors to look for savings in value engineering exercises. They expect the six-month lead-in to provide big savings because of this incentive.[7]

Quality control

Quality control is based on self-certification by the contractor, which Rail Link Engineering believes is consistent with partnering. Rail Link Engineering has approved the contractor's quality systems, monitors that they are being applied correctly, and does spot checks on the finished work. An essential aspect of self-certification is that defects identified by the contractor are paid for as part of the actual cost. However, if defects get right through a contractor's quality control system to Rail Link Engineering's inspection of the finished work, they are not paid for. This is intended to provide a real incentive for contractors to ensure that their quality control procedures are applied consistently.

Overall performance

To date, there is every indication that Section 1 of the project will be completed on time and under budget. The same team has the chance to build on the lessons that Section 1 provides, so there is every reason to expect that the much more demanding Section 2 will also meet its targets.

A good indication of the partnering attitudes that have developed throughout this massive project is that, to date, it has not been necessary to go to adjudication on any of the contracts. This is a truly remarkable testament to the way this huge and important project is being managed.

Channel Tunnel Rail Link: Ashford contract

Contract 430 is the largest of the individual contracts on Section 1 of this project, with a value of £150 million. It involves three distinct sectors of railway. The first is a straightforward new line through green fields. The second sector is in a cut-and-fill tunnel, and the third is 70% viaduct and 30% through old industrial sites. The work involves many interactions with existing roads and railways. From the outset the aim has been to complete on time and to stay within the overall budget. Yet projects of this scale and complexity in the UK traditionally go wrong, nearly always over-running on cost and time.

The project manager for the Rail Link Engineering consortium, which is responsible for design and project management on behalf Union Railways South, clearly understood from day one that if an established traditional approach were used, the project would almost certainly not meet its objectives. So he decided to find a better

---

7 Authors' note: This only applies in the target contracts, because the target remains the same, with the potential for lower total Actual Cost resulting in a greater share of savings for the contractor.

way of working, and this determination was a crucial starting point for a remarkable and ongoing story of co-operative teamworking.

Appointment of the contractor

Bids were invited in 1997 for the construction on the basis of an outline design. The bid documents stated that NEC, Option C, Target Contract with Activity Schedules, would be used. They stated Rail Link Engineering's intention to run the contract on a partnering basis, and initially indicated a six-month period after bid evaluation to allow the selected contractor to understand the project before work started on site.

Tenderers were required to state how they would deal with the contract, including their proposals for partnering. Evaluation was based on price and technical and partnering competence. The contract was awarded to Skanska, who in the bid presentation and interview emphasised key features of their approach as being:

- an open book basis which gave effect to NEC procedures,

- a non-adversarial approach,

- resolving problems at project level,

- respecting roles but not allowing them to create barriers to the best ways of working, and

- putting the project first in the belief that a successful outcome benefits everyone.

Initial project team

Skanska's first task was to understand the current state of the Works Information and approvals and permissions from local and statutory authorities, because design had continued during the bidding process. They found the design was 35–40% complete, and that the approvals regime was complex and tough. Railtrack in particular have been very demanding on environmental issues as well as railway issues. Whilst the bid documents had promised a six-month period for this initial evaluation, restructuring of the project reduced this to one month.

The project team therefore had to make a very fast start. This was achieved by the whole project team, then about 12 people from Rail Link Engineering and Skanska, moving into an office in London that happened to be vacant and available. On day one, the team members sat wherever there was a desk and got on with the enormous task of understanding the difficult construction project they had taken on. They worked incredibly hard to produce an overall plan for the project based on the outline design that was reflected in an agreed target cost that formed the basis of the contract.

The intensive effort left no time to think about the distinct roles of the consultant and contractor. The team just got on with whatever needed doing. Thus began a distinctive approach based on co-operative teamworking, but it had to be given constant attention as the team built up to its total strength of just over 1000 people (nearly 840 involved in the direct construction work and about 160 in the project office).

The build up of the target cost is recorded in a detailed document, which once it was agreed was never looked at again. This is because under NEC target contracts, the build up of the target cost is not used in determining whether a compensation event has arisen. While the question of whether all the effort devoted to producing it could not be used more productively might be raised, the detailed documentation was necessary to evaluate that the contractor had fully understood the contract and that the target price was robust.

The project team at Ashford

Work began in 1998 with the establishment of a project office at Ashford and the establishment of work sites. Joint teamworking had already developed naturally from the pressure of work and the informality of the accommodation in the team's temporary London office. This spirit was deliberately carried over into the project office on site. The office is open plan and arranged according to the logic of the work, not on the basis of allegiance to firms. The client, Rail Link Engineering, Skanska and key subcontractors work together as one team.

Subcontractors are employed using ECS, Option C subcontracts and are fully brought into the partnering team.

There is a clear understanding that achieving outstanding performance demands constant attention throughout the team. As Rail Link Engineering's contract manager said: 'It is easy to agree mutual objectives, but finding the best way of achieving them has to be worked at constantly'.

All Rail Link Engineering's and Skanska's senior managers on site now understand that the right priorities for the project are safety, quality and time. They have learned that if they get all three right, then they have a realistic chance of achieving high efficiency, which delivers low costs. This mature approach is central to their growing success.

An important effect of the common project team office that has helped build team spirit is that people do not make destructive comments about each other. Irritations and problems are expressed, often very directly, but since they are likely to be challenged in the open office, views expressed have to be capable of being justified. Not everyone has changed; there remain people with negative attitudes who are waiting for things to go wrong. However, their criticisms of ideas sometimes help identify real weaknesses. This is leading to an implicit recognition of the need for effective teams to include a broad spread of personalities. Although there have been no conscious efforts aimed at forming balanced teams, the team has begun to understand that it needs to accommodate different, even conflicting, points of view if it is to make good decisions.

In addition to the project office there are site offices for each of the three main sectors of the work that house joint consultant-contractor teams that are responsible for day-to-day construction.

There is a weekly meeting of the site's senior managers to resolve any problems. The general aim is for problems to be resolved within two weeks at the work level. If this is not possible, the problem is referred to the weekly meeting. This rarely happens. If the weekly meeting were unable to resolve a problem, the two senior figures, Rail Link Engineering's and Skanska's contract managers, have agreed that they will deal with it.

Partnering

Partnering was given a positive boost by a two-day workshop held about one month after the team was established at Ashford. This was run by a professional facilitator from the John Carlisle Partnership, and was attended by the senior people in the team from the client, Rail Link Engineering and contractor. By the time the workshop was held the team understood the great difficulty of the project and the challenges they faced, and so were ready to co-operate in looking for a better approach.

North Downs tunnel

The workshop began with discussions and games aimed at understanding each other and establishing the value of co-operation. Then, groups discussed the outcomes they wanted and how they expected the project to be run.

The results from the workshop were embodied in a partnering charter, which includes a project logo and an agreed mission statement. This has been updated by subsequent workshops and the current version is pinned to the walls throughout the project and site offices. The most important results were removing peoples' doubts about senior management's commitment to co-operative working, and raising everyone's expectations of how other people would behave.

As more people came onto site, further workshops were held, so that everyone in the project office attended a workshop and had the opportunity to contribute to the current partnering charter.

The participants at the first workshop were brought together for another two-day workshop 10 months later. This served to define the team's mature approach of recognising the need for a balanced team that is enthusiastic and positive, but which can accept and use sceptics creatively to challenge ideas and ensure that potential problems are not missed. The workshop helped the project's senior managers see partnering, like a marriage, as being dependent on trust, openness and honesty. They also recognised that personal relationships between the people at the top, what they call the 'chemistry' between the key people, is crucially important.

Design information

Design information is produced in the Tottenham Court Road office of the overall Channel Tunnel Rail Link project team. Rail Link Engineering's contract manager provides formal liaison between the design office and site. Design meetings take place on site every two weeks, which bring together Rail Link Engineering's design managers and key members of the Ashford project team.

The speed of working has prevented close integration of design and construction expertise. Also, formal consents and permissions take anything up to 12 weeks to obtain. In traditional practice these factors would have resulted in design meetings starting with a programme designed to highlight the delays and disruption being caused by late design information. The team spirit at Ashford enabled the meetings to recognise quickly that they needed to establish the construction priorities for design information, so that what was really needed was available as early as possible.

Quality control

Quality control is based on self-certification by the contractor and subcontractors. Skanska, especially, has a well-established quality management system, built up from experience of PFI and design and construct projects, which required little adjustment to fit the NEC model.

The system in use at Ashford requires the workers undertaking the construction work to do their own quality checking. Skanska has a team of inspectors who make random checks and sign off concrete pours and other major activities. They report directly to Skanska's senior manager on site. This is intended to ensure that everyone involved in construction takes quality seriously. The importance of quality is further reinforced by carrying out the required tests on concrete and earthworks in a site-based laboratory, to avoid delays in getting the results.

Rail Link Engineering approved the quality systems at the start of the project and, as construction work proceeds, carry out spot checks on site. For example, they test 10–20% of concrete pours. The client has approved the quality systems and there are occasional inspections on site by central government inspectors to ensure the quality of the work. By any standards, there is a real belt-and-braces approach to quality control and there have been no significant quality problems.

Time control

The number of changes has made it difficult to stick to the original completion date. The ECC compensation event procedures, which require the time effects of each event to be calculated, can result in a major re-programming exercise for each event. The team works together to minimise the effect of the changes and to streamline the ECC procedures. This has required a lot of hard thinking by the whole team.

Frequent workshops are held to find ways of keeping the work on programme. These review the current definition of the required work and look at design and construction options and the cost consequences. Then decisions are made by the team. In this way the initial construction method for the main tunnel in the project, which used one set of props that were progressively moved down the tunnel, has been changed. By buying additional props and doubling up on other key resources, two workfaces were opened up instead of one. The extra costs to the contractor were less than the delay damages for late completion.

The key meeting in maintaining control is the weekly progress meeting held from 4.00 to 7.00 pm every Monday. It is attended by the sector managers and contract managers. A formal Weekly Progress Report guides the meeting, which is crucial to the management of the project. The meeting is purposeful, in no small part because

The Medway Bridge

the formal document is issued at 10.00 am on Mondays, so that everyone has the chance to read it before the meeting, do their homework and be ready to deal with the main matters raised by the report.

The Weekly Progress Report describes the progress achieved the previous week and states the current plan for the next two weeks. The document concentrates on agreed key performance indicators, dealing with safety, quality and productivity. It makes extensive use of graphical representations of trends to concentrate attention on the overall picture.

Each sector and each main type of work is looked at in turn. The report includes any issues or problems that anyone wants to have discussed. As well as the graphical summaries, the Weekly Progress Report contains detailed information provided and agreed by the sector managers, and so is treated as a factual record of what has happened in the previous week. This is very important because it avoids arguments over the facts. It is accepted that the Weekly Progress Report records the facts, and it is taken as a serious matter if anyone needs to dispute what it reports. Similarly, it is not acceptable to raise issues or problems not recorded in the formal report. This discipline allows people to prepare for the meeting, and then at the meeting to concentrate on finding the best answers.

It took a lot of work and consistent effort from senior managers to get everyone to agree to, and then accept, the one record of progress. There are great temptations in traditional practice to keep personal records and not to tell other parties the full story. These traditional attitudes had to be overcome to get a clear, accurate and accepted single record of progress.

The project's senior managers do not just rely on the formal Weekly Progress Report; they walk the job at least once, and more usually twice, a week. They talk to people on site about their own progress and ask questions about any problems they see. This is an important two-way process aimed at ensuring that the project stays on track and everyone on site knows they have an important part to play in their joint success.

Cost control

The target cost is adjusted by means of compensation events, most of which are now identified at early warning meetings. Initially it was difficult to stick to the NEC compensation event timetables, because of the number of changes resulting from the amount of design development needed after the contract was agreed. However, the team has come to the conclusion that it is worth the effort needed to stick to the NEC timetables because this forces decisions to be made quickly.

Another initial problem with the ECC procedures was that early warning meetings were used by the contractor's commercial people to concentrate on looking for compensation events. However, now their real purpose of solving problems on site has been established as a result of Rail Link Engineering's approach, they are seen as a powerful tool in ensuring that the project keeps moving fast.

Value engineering workshops were used early in the project to look for better construction ideas. These generated over 60 ideas for improving the construction method. One significant idea eliminated the need for tunnelling, and replaced it with cut-and-cover methods that saved time and made safety and costs more certain.

At 60% of the way through the project, there had been about 250 early warning meetings and just short of 100 compensation events. The time taken to assess compensation events has been reduced by not evaluating every new issue in great detail as it is only the target that it affected. The contractor is paid the actual cost.

Skanska are paid every four weeks their actual costs projected forward seven weeks, so that when payment is made all their costs to that date are covered. This reduces the contractor's costs in financing the project. Skanska setting up a separate bank account for the project has helped the calculation of actual costs. All expenditure is authorised in the project office and goes through the project bank account. Costs are allocated to some 2300 third-level programme activities. Auditing is carried out by the project manager, client and financiers, and so is thorough and virtually constant.

Procedures

The project team has developed a set of project procedures that describes how they have agreed to work. There are currently 28 procedures dealing with all the key activities. For example, one describes the preparation of the Weekly Progress Report, stating what matters are to be considered and who is to provide what information. Another describes the issue resolution process aimed at ensuring that the people directly involved resolve problems quickly.

The aim of each procedure is to have a short description of how the team has agreed to act. The procedures consist of short purposeful statements of actions to be taken by specified people. They take into account NEC procedures, firms' structures and procedures, and the individual characteristics of team members who have to apply the procedures.

The procedures are changed when a better way of working is agreed. They provide an effective means of introducing new people to the project, and help ensure that good ideas are not forgotten and practice does not become sloppy.

Outcomes

The project was on programme at the time of writing, and there was every expectation that it will exactly meet the completion date set in the contract. Indeed, progress has steadily improved, as measured against the original plans, and most early start dates are now being achieved, leaving the team the whole of the time float to deal with problems.

Total actual cost is slightly below the current target and is expected to remain so right to the completion. These costs are just below the overall budget for the Ashford project. One important implication is that Skanska are likely to achieve the profit level included in their original bid.

No significant problems have arisen with either safety or quality.

The most remarkable outcome has been the development of a real team drawn from different firms. The key people involved believe that NEC and partnering have played important roles in this success. They also believe that NEC and partnering reinforce and support each other and find it difficult to see how either can work properly without the other.

When the contract was about 50% complete, the senior managers on site from Rail Link Engineering and Skanska reviewed what had been achieved and considered how they could do even better in the second half. The plans produced from a series of meetings between the two contract managers take teamworking significantly further than anything so far achieved.

The future

The aim for the second half of the project is to create one seamless team led by Skanska's contract director with Rail Link Engineering's contract manager as his deputy. The team will work to one integrated programme with one goal. It will focus on the main issues and cut out unproductive administration. Effort will be concentrated only on work that really adds value; duplication will be cut out. For example, the effects of compensation events will be calculated and accepted. Quality control will be streamlined. Paperwork will be drastically reduced. The best person available from within the existing team will be appointed for each essential task, irrespective of which firm he or she was originally employed by.

The expected benefits are that the project office will be more effective and about 40 people will be released from the project team. This 25% reduction in the number of people in the project office is an interesting and important measure of the benefits of co-operative teamworking supported by NEC and partnering.

Picture acknowledgments

Courtesy of Halcrow Group Ltd

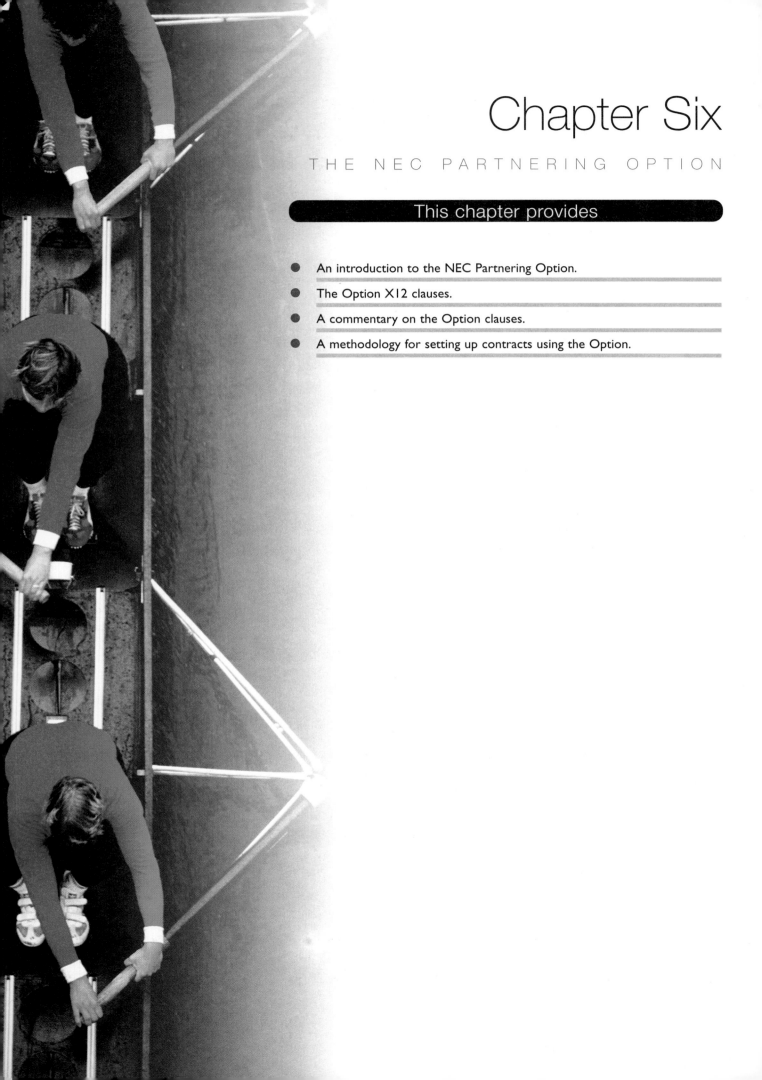

Chapter Six

This chapter provides

- An introduction to the NEC Partnering Option.

- The Option X12 clauses.

- A commentary on the Option clauses.

- A methodology for setting up contracts using the Option.

The NEC Partnering Option, published in June 2001 as Option X12, provides a simple and flexible way of adding a partnering basis to ECC, PSC and ECSC bi-party contracts. The introduction and Option X12 Clauses are reproduced here in full and are self-explanatory. The published Guidance Notes for the Clauses are amplified in the commentary, which follows.

Use of the Option requires the partnering organizations to complete the necessary Contract Data for the Option and produce three documents. These documents are the Partnering Information, the Schedules of Partners and the Schedule of Core Group Members. Partnering organisations then include the Option into the stated conditions of contract for their respective bi-party contracts.

A completed example for use with an ECC based contract of Contract Data and each of the three documents is provided in the methodology for setting up contracts using the Option.

Introduction to the NEC Partnering Option

A partnering contract, between two Parties only, is achieved by using a standard NEC contract. Option X12, which puts the NEC Partnering Option into a contract, is used for partnering between more than two parties working on the same project or programme of projects. The Partnering Option is used as a secondary option common to the contracts that each party has with the body that is paying for its work. The parties who have this Option included in their contracts are all of the bodies who are intended to make up the project team. The Partnering Option does not create a multi-party contract.

The option does not duplicate provisions of the appropriate existing contracts in the NEC family that will be used for the individual contracts. It follows normal NEC structure for an option, in that it is made up of clauses, data and information.

The content is derived from the Guide to Project Team Partnering published by the Construction Industry Council (CIC). The requirements of the CIC document that are not already in the NEC bi-party contracts are covered in this option. The structure of the NEC family of contracts means that this Partnering Option will not work unless an NEC contract is used.

The purpose of the Option is to establish the NEC family as an effective contract basis for multi-party partnering. As with all NEC documents, it is intended that the range of application of this document should be wide. By linking this document to appropriate bi-party contracts, it is intended that the NEC can be used for

● partnering for any number of projects (i.e. single project or multi-project),

● international projects,

● projects of any technical composition, and

● as far down the supply chain as required.

The option is given legal effect by including it in the appropriate bi-party contract by means of an additional clause (Option X12 in the ECC for example). This option is not a free standing contract but a part of each bi-party contract that is common to all contracts in a project team.

The underlying bi-party NEC contract will be for a contribution of any type, as contractor or consultant for example, the work content or objective of which is sufficiently defined to permit a conventional NEC contract to be signed. The content need not be well defined for a cost reimbursable or time based contract in the early stages.

Parties must recognise that by entering into a contract which includes Option X12 they will be undertaking responsibilities additional to those in the basic NEC contract.

A dispute (or difference) between Partners who do not have a contract between themselves is resolved by the Core Group. This is the Group which manages the conduct of the Partners in accordance with the Partnering Information. If the Core Group is unable to resolve the issue then it is resolved under the procedure of the Partners' Own Contracts, either directly or indirectly with the *Client*, who will always be involved at some stage in the contractual chain. The *Client* may seek to have the issues on all contracts dealt with simultaneously.

The Partnering Option does not include direct remedies between non-contracting Partners to recover losses suffered by one of them caused by a failure of the other. These remedies remain available in each Partner's Own Contract, but their existence will encourage the parties to compromise any differences that arise.

This applies at all levels in the supply chain, as a Contractor/Consultant who is a Partner retains the responsibility for actions of a subcontractor/subconsultant who is a Partner.

The final sanction against any Partner who fails to act as stated in the Partnering Option is for the Partner who employed them not to invite them to partner again.

There are many scenarios in setting up a project. The NEC family of contracts with the Partnering Option is sufficiently flexible to deal with them. For example, the contract may be a Professional Services Contract with a Contractor for a feasibility study. Subsequently that Contractor may do later work using the Engineering and Construction Contract. The Contractor may be a Partner during both stages of his contribution to the project.

Additional Contract Data for the Option

The *Client* is the Party for whom the projects are being carried out. He may also be the Employer in an NEC contract.

The *Client's* objective is the objective for the 'programme of projects' if more than one or for 'the project' if only one. The objective should be expressed quantitatively if possible (the business case). It should also include the partnering objectives.

Partnering Information includes any requirements for

- use of common information systems, sharing of offices,

- attendance at Partners' and Core Group meetings,

- participation in partnering workshops,

- arrangements for joint design development,

- value engineering and value management,

- risk management, and

- other matters that the Core Group manages.

This information should not duplicate requirements in the bi-party contracts.

The additional contract data for the Option, like other Contract Data in the NEC contracts, does not change. The Schedule of Partners and the Schedule of Core Group Members, like the Activity Schedule and other schedules referred to in Contract Data do change from time to time. Samples of the typical information required in the Option Data and the schedules are provided in a later section of this Chapter.

The Conditions: Option X12: Partnering

These conditions are incorporated into an ECC or PSC contract by including them in the list of *conditions of contract* in the first statement of Contract Data provided by the *Employer*. They are included into an ECSC contract as additional conditions of contract at the end of the Contract Data section.

Actions X12.1

(1) Each Partner works with the other Partners to achieve the *Client's* objective stated in the Contract Data and the objectives of every other Partner stated in the Schedule of Partners.

(2) Each Partner nominates a representative to act for it in dealings with other Partners.

(3) The Core Group acts and takes decisions on behalf of the Partners on those matters stated in the Partnering Information.

(4) The Partners select the members of the Core Group. The Core Group decides how they will work and decides the dates when each member joins and leaves the Core Group. The *Client's* representative leads the Core Group unless stated otherwise in the Partnering Information.

(5) The Core Group keeps the Schedule of Core Group Members and the Schedule of Partners up to date and issues copies of them to the Partners each time either is revised.

(6) This option does not create a legal partnership between Partners who are not one of the Parties in this contract.

Identified and defined terms X12.2

(1) The Partners are those named in the Schedule of Partners. The *Client* is a Partner.

(2) An Own Contract is a contract between two Partners which includes this option.

(3) The Core Group comprises the Partners listed in the Schedule of Core Group Members.

(4) Partnering Information is information which specifies how the Partners work together and is either in the documents which the Contract Data states it is in or in an instruction given in accordance with the contract.

(5) A Key Performance Indicator is an aspect of performance for which a target is stated in the Schedule of Partners.

Working together X12.3

(1) The Partners work together as stated in the Partnering Information and in a spirit of mutual trust and cooperation.

(2) A Partner may ask another Partner to provide information that he needs to carry out the work in his Own Contract and the other Partner provides it.

(3) Each Partner gives an early warning to the other Partners when he becomes aware of any matter that could affect the achievement of another Partner's objectives stated in the Schedule of Partners .

(4) The Partners use common information systems as set out in the Partnering Information.

(5) A Partner implements a decision of the Core Group by issuing instructions in accordance with its Own Contract.

(6) The Core Group may give an instruction to the Partners to change the Partnering Information. Each such change to the Partnering Information is a compensation event which may lead to reduced Prices.

(7) The Core Group prepares and maintains a timetable showing the proposed timing of the contributions of the Partners. The Core Group issues a copy of the timetable to the Partners each time it is revised. A Partner incorporates information in the timetable into its Own Contract programme.

(8) A Partner gives advice, information and opinion to the Core Group and to other Partners when asked to do so by the Core Group. This advice, information and opinion relates to work that the other Partner is carrying out under its Own Contract and is

given fully, openly and objectively. The Partners show contingency and risk allowances in information about costs, prices and timing for future work.

(9) A Partner notifies the Core Group before subcontracting any work. A Partner is responsible under its Own Contract for the actions and inactions of its subcontractor.

Incentives X12.4

(1) A Partner is paid the amount stated in the Schedule of Partners if the target stated for a Key Performance Indicator is improved upon or achieved. Payment of the amount is due when the target has been improved upon or achieved and is made as part of the amount due in the Partner's Own Contract.

(2) The *Client* may add a Key Performance Indicator or associated payment but may not delete or reduce a payment stated in the Schedule of Partners.

Commentary on the Clauses

The NEC Panel have provided an Option that binds the partners to the *Client's* and each other's objectives by being an obligation of their Own Contract, rather than develop a completely new multi party partnering contract. The arguments for and against the NEC 'option' approach, versus use of a purpose designed multi-party partnering contract, will doubtless be debated in many forums and for many years to come.

The advantages for the NEC approach would seem to be

- Avoidance of yet another contract, with its high cost of learning and initial application. Those wishing to try partnering who already use NEC will not need to learn a new contract; however they will need to understand and practice the correct use of NEC (See Chapter 2) and make the necessary culture shift to which this Guide has made repeated reference.

- Greater time and scope of work flexibility for introduction of new members to the partnering team than might be available if all partners had to sign up to a single multi party partnering agreement from the outset.

- Contractual arrangements remain consistent up and down the procurement chain, as well as across the partners. Contracts are still the same with organisations who are not required to be part of the partnering team (except that Option X 12 would not be included of course), which they would not be if a multi party partnering contract were to be used. Avoiding conflicting obligations between contracts within the same project is a vital contribution to the success of any project, whether partnering or not.

- Complex contractual liabilities between all Partners are avoided with the benefit that information is likely to be given more freely, thus fostering the partnering process.

Liability of one Partner to another

A full explanation of possible duties of care and liabilities such as tort, which may exist between partners who are not in contract with each other, is beyond the scope of this Guide. As the issue of liability differs from one jurisdiction to another, readers should consult the necessary expertise applicable to the law governing the contract they wish to develop. Partnering arrangements through contracts, whether in multi-party form or through the NEC Option approach, are relatively new. The implication of The Contracts (Rights of Third Parties) Act 1999 will need to be considered for contracts entered into in terms of UK law. Similar implications arising from either statutory or common law will apply in other jurisdictions.

In the NEC Option, each Partner is made responsible for the actions of its subcontractor. Hence liabilities exist through each link in the contractual chain, which is the normal situation.

Notes about the clauses in Option X12

The following notes are not intended to be a substitute for the definitive guidance provided with the published NEC Partnering Option, but merely further commentary, which may be helpful to users. They should be read together with the published Guidance Notes and the Option X12 Conditions.

Clause X12.1 Actions of Partners and the Core Group

(1)   Although clause X12.1(1) states that Partners only have a contractual commitment when there are two or more Own Contracts with the *Client* in existence, it is likely that the X12 Partnering Option will be drafted and put into effect informally well before the formal contracts come into existence. Firms often work together before entering into formal contracts. In particular, firms often identify the Partners, decide who will form the Core Group and agree the Partnering Information before entering into formal contracts. In strategic partnering where the partnering arrangement continues over more than one project, these details are almost always agreed outside of any one project. Such agreements often result in considerable work being undertaken to identify the best ways of working together. This informal partnering does not alter the formal contractual position once a contract dealing with one project is signed.

(2)   Noting that all references in NEC to the singular also mean the plural, it may be appropriate in terms of clause X12.1(2) for some Partners to nominate more than one representative because they have several important interests that should be taken into account in making decisions. Obvious examples include a client being represented by a finance director, a facilities manager, and a representative of the people who will use the new facility; and a design build contractor being represented by a lead designer and a construction manager. In these cases all the representatives are listed in the Schedule of Partners. One of the representatives may be nominated to act as the formal representative but this is not essential. It is often more efficient for the views of all the important interests to have an equal voice in decision making.

(3)   As well as acting on matters stated in the Partnering Information, it is inevitable that the Core Group will also have a substantial hand in managing the project, albeit at a strategic level through all the Partners and their Own Contracts, except in situations where the Core Group may be named in the ECC as the *Project Manager*.

(4)   The second sentence of clause X12.1(4) provides the default position that the *Client*'s representative leads the Core Group. It often makes sense however, for someone other than the *Client*'s representative to chair meetings of the Core Group as the case studies illustrate. Indeed it often makes sense for different people to chair meetings depending on the particular subject under discussion. The arrangements decided on should be stated in the Partnering Information.

(5)   There are a number of administrative actions of the Core Group such as the one stated in clause X12.1(5) and the Core Group will need to decide at the outset who will undertake and be remunerated for these on behalf of the Group.

(6)   Clearly there is a legal relationship between the parties to each Own Contract. Clause X12.1(6) addresses the legal status of Partners who are not in contract with each other and of all Partners acting as a group. (See the notes above concerning liability between Partners.)

Clause X12.2 Identified and defined terms

(1)   The second sentence of clause X12.2(1) confirms that the *Client* is always a Partner. This may seem obvious from a control point of view, but further stresses the importance of the client in any project situation. Decisions need to be made quickly and decisively in most if not all projects and clients cannot simply abdicate their role to the professional team and then complain afterwards that they did not receive the outputs they expected. It also focuses on the need for the required delegated authority within the client body.

(2)   An Own Contract is not always one that has the *Client* as one of the parties. An Own Contract between main contractor and a subcontractor, or between consultant designer and a specialist sub-consultant will be required if either the subcontractor or designer is to become one of the Partners.

(3)   The members of the Core Group, defined in clause X12.2(3) as the Partners listed in the Schedule of Core Group Members, should be selected carefully to ensure that the Core Group includes everyone likely to contribute to the overall leadership of the project. Mature partnering goes beyond approaches based on managers issuing instructions and instead relies on competent teams cooperating in making decisions aimed at agreed objectives. There remains a need for leadership to help teams achieve the best results. This should be the role of the Core Group, which would revert to a traditional project management role only where the teams are failing to perform acceptably. As the case studies illustrate, firms experienced in partnering may decide to name the Core Group as the *Project Manager* in ECC Own Contracts to give expression to this mature approach.

(4)   Clause X12.2(4) makes the point that Partnering Information may be in an instruction given in accordance with the contract. This implies that Partnering Information may change and Clause X12.3(6) confirms that such a change is a compensation event, obviously under each Own Contract. Whilst it should be obvious that such instruction is in respect to the Partnering Information, it would be advisable for users to make a clear distinction between instructions given in accordance with the contract where they relate to Works Information or the Scope, and Partnering Information to avoid possible confusion later.

(5)   It is important in establishing Key Performance Indicators and targets, defined in clause X12.2(5), to ensure that they do not create conflicting interests. Everyone employed on any one project should be working towards exactly the same overall targets. Where this is not the case, conflicts are almost inevitable and this of course tends to undermine partnering. It needs a very mature approach to partnering to achieve completely integrated targets, one practical approach is illustrated in the example given later in this chapter.

It is particularly important in establishing Key Performance Indicators and targets for subcontractors to ensure that individual targets do not create conflicting interests for the main contractor and subcontractor. The same considerations apply to Key Performance Indicators and targets for subconsultants.

Clause X12.3 Working together

(1)   Clause X12.3(1) requires the Partners to work together as stated in the Partnering Information and in a spirit of mutual trust and cooperation, irrespective of whether they have a contract between them or not. This

obligation is already in each Own Contract. Much has been said, not always very complimentary, about this obligation, introduced into contracts as a result of the Latham report[1]. The main observation is that when things go wrong, mutual trust and co-operation go out of the window. This is the justification for having a neutral partnering advisor, whose main function is to encourage the parties to behave as stated, even when difficulties arise. Workshops during the course of a project are used and team social events play their part as well. For those using NEC as it is intended to be used, the obligation is very easy to identify with as it works well for them.

(2)   Clause X12.3(2) places obligations on Partners for the exchange of information which another Partner may need to carry out his own work. Partners would be advised when doing so to address the matter of liability for the given information in the exchange documentation. Problems may arise when such an exchange is not possible, without prior arrangement, due to constraints on the giving Partner in respect to the transfer of rights, particularly intellectual property. The intent of the clause is fine, but clearly there is more behind it than meets the eye. Mature Partners will no doubt understand the constraints and liabilities and deal with such formalities as part of their everyday business.

(3)   Clause X12.3(3) further emphasises the partnering ethos by requiring a Partner to give early warning as soon as he is aware that another Partner's objective could be adversely affected. The concept of early warning as used in NEC contracts is probably one of its best features. The procedure for dealing with early warnings is given in each Own Contract but is not given in Option X12 for early warning between Partners not linked by Own Contracts. In these cases it would be up to the Core Group to decide how such early warnings would be dealt with, and any particular methodology should be included in the Partnering Information.

(4)   The actual extent of the common information systems required in terms of clause X12.3(4) is set out in the Partnering Information. This should include whether this requirement applies to planning and programming as well.

(5)   The procedure for how the Core Group acts is stated in the Partnering Information as required by clause X12.1(3). When the actions of the Core Group result in decisions to be carried out, clause X12.3(5) requires that the Partners carry them out by issuing instructions in terms of their Own Contracts. The terms of each Own Contract will then deal with the effect of the instruction, including whether it is a compensation event.

(6)   Because clause X12.3(6) confirms that when the Partnering Information is changed, it is a compensation event (under the affected Own Contract) users will need to consider carefully how they prepare the Partnering Information, avoiding too much minor detail wherever possible. It is also one of the compensation events which may lead to reduced Prices in the Own Contract.

(7)   Clause X12.3(7) requires the Core Group to prepare and maintain a timetable showing the proposed timing of the contributions of the Partners. Each ECC and PSC Own Contact requires detailed programmes that show dates when the Employer and Others (a defined term) are required to provide information and acceptances. In a partnering context, Others will include other Partners. Hence

1 *Sir Michael Latham,* *Constructing the Team, Final Report of the Government / Industry Review of Procurement and Contractual Arrangements in the UK Construction Industry,* The Stationery Office, London July 1994

the timetable which the Core Group prepares, in a project, is merely a co-ordinated review of all interface dates which provide the inputs for programmes required in terms of Partners Own Contracts and other contracts not involving Partners. The timetable simply provides the links between Own Contracts and so may be implicit rather than needing to be a separate additional document. In the case of a strategic alliance designed to handle a number of projects, it will be separate high level document used to optimise Partners available resources.

(8)   Clause X12.3(8) requires Partners to give advice, information and opinion to the Core Group at any time, and to other Partners when asked to by the Core Group. This is the contractual obligation inserted to formalise rights to open exchange of information. However, mature partnering encourages everyone involved to take responsibility for the overall performance of the project and if they see a problem or have a good idea, they should give their advice, information and opinion to the Core Group. The same constraints surrounding the exchange, as pointed out in (2) above, will require sensible protocols to be established between the Partners from an early stage in the project or the alliance.

(9)   Clause X12.3 (9) requires that the Core Group be informed before a Partner subcontracts any work. This is in addition to constraints regarding subcontracting already in the Own Contract. No procedures arising from this notification are provided, although there are some suggested procedures in the Guidance Notes. The purpose of this clause is to allow the Core Group to exercise the necessary overview, which is its function, and provide an opportunity to decide, for example, whether or not the proposed subcontractor (or subconsultant) should be included as a Partner, or perhaps not used at all.

Clause X12.4 Incentives

(1)   Clause X12.4(1) deals with payment of incentives when stated targets have been achieved or improved upon. Users are advised to consider targets with great care. Amounts stated in the Schedule of Partners should be linked to clear statements in the Partnering Information of just how and by whom the achievement of each target is to be established, and the amount due assessed. The outcome should be capable of adjudication under each Own Contract. Emotions over targets and their payments have destroyed many a good working relationship. The Guidance Notes infer that targets should be set such that if a target is not met for any reason, the payment for achievement is not made and if one Partner lets the others down by poor performance, all lose their incentive payment. These are tough concepts, which require a mature approach. Particular care is required when one of the NEC target contracts (usually option C) is being used.

(2)   Clause X12.4(2) provides the *Client* with the authority to add a Key Performance Indicator (KPI). Although the Guidance Notes state that the *Client* should consult the other Partners before adding a Key Performance Indicator, this is not a requirement of the Option X12 clause. The *Client* will need to take account of the effect on all parties in the procurement chain before adding another KPI. The Environment Agency case study in Chapter 5 points out very clearly how targets set for a contractor, when passed down back-to-back to his consultant, had the opposite effect on the ability of the consultant to achieve his target.

(3)   The authors are of the opinion that targets should remain high level and strategic in nature rather than be related to detail aspects of work to be undertaken. There could be too much opportunity for manipulation or 'taking the eye off the ball' if targets are related to work activities that form a relatively small component part of a total programme of work.

Methodology for setting up contracts using Option X12: Partnering

A *Client*, (also a Partner) possibly using a partnering advisor, having decided to use a partnering arrangement will prepare the Option Data (Additional Contract Data, with *Client* details, *Client's* objectives and Partnering Information) and the first Schedules of Partners and Core Group members.

A completed example

The following completed example of using Option X12 in an ECC based contract illustrates the description of mature partnering given in Chapter 4. It relates to a hypothetical project in which the *Client's* business case sets a limit to expenditure.

The *Client* has decided to carry the burden of unforeseen financial risks in order to enable the other Partners to concentrate on producing the best value within the stated budget.

The *Client* has guaranteed the other Partners an agreed profit. Therefore there are no additional payments to the other Partners for achieving the targets, as the certainty of making an agreed profit is judged to provide sufficient incentive.

Other customers, perhaps with less experience of working with the other parties, may want to use incentives that require savings and cost over-runs to be shared. Such arrangements need to be carefully designed so that all the firms involved are working towards the same targets. This is important because incentives that cause team members to be working towards different targets almost inevitably cause conflicts and undermine partnering attitudes. Therefore the following example is based on approaches used in developed, mature partnering and so provides a guide to best practice.

New office building for Major Supermarket plc;

At the outset of agreeing to partner, the main contributing Partners will prepare the Option Data and Schedules for future inclusion into each Own Contract (either ECC or PSC) as explained later in this Chapter.

Option Data for this example could be as follows:

Option X12: Partnering

The *Client* is Major Supermarket plc

Address   Food Hall House, Cheapside, London EX5 7GT

Contact Details John Brown, Chief Facilities Manager.

Telephone: 020 7568 9008; Fax: 020 7568 9108

E-mail   johnbrown@majorsuper.com

The *Client's* objective is:

To move the 520 staff of Major Supermarket plc Property and Construction Division into a new office building on their site adjacent to Food Hall House at 12.00 noon on 1 March 2002. The new office building is to provide space that is at least of the same quality and standards of comfort as that currently occupied by the Property and Construction Division. The total budget for the design, construction, furnishing and equipping of the new office building, ready for the staff to move into, is £8m.

To ensure that the other Partners achieve their agreed financial objectives, which include an agreed profit. The *Client* will carry the risk of cost and time over-runs with the intention that this will allow the other Partners to concentrate on ensuring that the *Client's* objective is achieved.

The Partnering Information[2] indicator is as follows:

The Partners and their essential support staff are based in a common project office in Room G17 in Food Hall House. Other staff are based elsewhere when the Partners agree this is in the best interests of the project.

All project meetings take place in the common project office. The *Client's* Project Manager will chair the meetings and his firm will be responsible for recording all meetings and issuing communications from the Core Group. Decisions are recorded on a white board as they are agreed. Everyone present is given a copy of the record of decisions. The record of decisions remains on open view in the common project office until the white board is needed for another meeting.

All project information that needs to be communicated other than through white board decisions is communicated in an electronic form agreed by the partners.

The management of the project is guided by formal meetings which include the following steps:

- everyone is introduced, if there is a new person present everyone should ensure they are clearly introduced,

- check what everyone wants to achieve from the meeting,

- ensure there is an agreed and clear agenda,

- ensure that all decisions and actions are recorded and agreed at the meeting so there is no need to circulate paperwork,

- close-down by checking how everyone feels about the meeting; and whether they have achieved what they wanted.

Project Process

Prior to the start of the project there is a Proposal Stage, which provides the business case for the decision to go ahead with the project.

The project starts with a Planning Stage and the Core Group being appointed by the *Client*. The project is managed by the Core Group through all subsequent stages to completion. They agree the scope, which should be consistent with the Partners' objectives, and establish a risk register for use by all Partners and for monitoring by the Core Group. Then the Core Group move to a creative decision making stage which begins with the first workshop.

The stage ends with a Baseline Document that records the agreed technical scope of work and quality standards, and a Project Strategic Plan that states how the work is to be carried out. The objective is to finalise Own Contracts for the Core Group and other Partners on the basis of the Baseline Document and Project Strategic Plan.

The Core Group include in the Project Strategic Plan recommendations for dealing with commercial and risk matters which affect all members of the project team and their respective contracts. These will include:

- identification of stakeholders, industry and other market related constraints, and long delivery time items,

- persons (firm) best able to develop contract documents,

- basis of selection of other contractors and suppliers,

- selection of the *Adjudicator*,

- typical secondary options to be used in ECC and PSC,

- ECC Section 8 risk allocation and additional conditions regarding limitation of liability and force majeure,

---

2 *Authors' note: The information given in this example is freely drawn from case studies of good practice described in chapters 3 and 5, and elsewhere, including particularly the well-developed approach used by Distribution Partnering described in chapter 3 page 25 of The Seven Pillars of Partnering.*

- latent defects,

- provision of Bonds and Guarantees,

- approach to quality and health and safety standards,

- additional compensation events,

- whole project insurance,

- professional indemnity and product liability insurance,

- transfer or otherwise of intellectual property rights,

- adverse weather data and selection of methodology within the terms of each ECC or ECSC contract being used,

- ground risk and dealing with unforeseen physical obstructions,

- Third Party consents and stakeholder/third party interference,

- dealing with under performance, default or insolvency of a Partner.

Some of the above matters will lead to the development of additional or special conditions of contract, which may be used in some contracts.

The Delivery Stage work is planned in detail and managed in a controlled manner in accordance with the plan using NEC procedures. The Core Group is expanded to bring in all those needed to complete the design and procurement, undertake the construction on site, commission the new facility ready for handover and produce the users' manual. It deals with problems and changes in a positive manner by setting up Task Forces to look for answers that allow completion without compromising performance or quality, on time, and within the budget. The Stage ends with the handover of the completed facility with zero defects, putting right any defects that have occurred, and a final partnering workshop to provide feedback for future projects.

The following meetings are used to manage the project:

Project Meeting

All Core Group members meet at 14:00 every Monday to deal with progress, cost, procurement, information flow and external influences.

Design Meetings

All Core Group members meet at 08.00 every Monday to deal with the design, any feedback from the customer; and explore alternative solutions.

Work Package Design and Procurement Meetings

All Core Group members and relevant Partners with an involvement in the work package meet when necessary to agree the exact scope of the work package, discuss and finalise the contract strategy and options to be used, and carry out detail checks of the contract documentation.

Work Package Meetings

Weekly meeting of construction managers with each specialist contractor's contract manager and site supervisor to deal with progress, detail design, construction plans and construction.

Work Package Interface Meetings

For the following major elements:

- groundworks,
- envelope,
- services,
- finishes,
- furniture and equipment.

Weekly meeting of all Core Group members and the specialist contractors involved in the work packages to ensure all information is in place for design coordination and efficient construction; to look for better ways of working; and resolve problems.

These meetings will be run as combined NEC early warning and compensation event meetings. Emphasis is to be given to alternative ways of doing things to avoid the need for extensions of time.

Site Meeting

Daily meeting over lunch of all Core Group members and site supervisors currently based on site to review today's progress; solve any problems affecting today's progress; and plan the next day's work. Review health and safety situation and take corrective action as required.

Partnering Workshops

All Partners attend partnering workshops. These take two days, are held at a neutral venue and are facilitated by an independent partnering facilitator appointed by the Core Group.

At the first partnering workshop, the Partners undertake teambuilding activities, review the Partnering Option Data and make any agreed changes, subject the initial brief to value management, identify risks, and agree the project objectives through the Baseline Document and the Project Strategic Plan.

Follow up workshops monitor progress and resolve problems. The final workshop:

- evaluates project performance compared to the original objectives,

- discusses problems and defects, and their causes,

- identifies good ideas that emerged from the project and considers how they contributed to the project's success,

- feeds the key results to all the Partners' senior managers.

Schedules of Partners and Core Group members

The following are examples of the typical information required in these schedules.

Schedule of Partners   Date of last revision: 25 January 2001

The Partners are the *Client* and the following:

Name of Partner	Management Consultants plc	Building Design Associates
Representative's Address and	Doug Jones, Management Team Leader	Mary Smith, Design Team Leader
Contact Details	Management House, Planning Way, London EA8 7PM, Telephone: 020 7765 5016, Fax: 020 7765 5016	Architect House, Design Way, London EA8 8AE, Telephone: 020 7890 2016, Fax: 020 7890 2216
	E-mail dougjones@managementcons.com	E-mail marysmith@buildingdesign.com
Contribution:	To provide time and cost control skills in cooperation with other members of the project team	To provide architectural design skills in cooperation with other members of the project team
Objective:	To earn a profit of £40,000 and ensure that all partners, including the client, meet their objectives.	To earn a profit of £70,000 and ensure that all partners, including the client, meet their objectives.
Joining date:	8 January 2001	8 January 2001
Leaving date:	1 March 2002	1 March 2002
Key Performance Indicator	Space equal to the quality and standards of comfort in the accommodation currently occupied by the Client's Property and Construction Division.	Space equal to the quality and standards of comfort in the accommodation currently occupied by the Client's Property and Construction Division.
Targets	Completion by 1 March 2002.	Completion by 1 March 2002.
	Completion for less than £8m.	Completion for less than £8m
	Zero defects at hand over.	Zero defects at hand over.
Measurement arrangements:	The *Supervisor* will evaluate the standards and quality of the space.	The *Supervisor* will evaluate the standards and quality of the space.
	Time will be managed and Completion will be determined in accordance with the provisions of the bi-party NEC contracts that incorporate this Partnering Agreement.	Time will be managed and Completion will be determined in accordance with the provisions of the bi-party NEC contracts that incorporate this Partnering Agreement.
	Cost will be managed and the final amounts due will be determined in accordance with the provisions of the bi-party NEC contracts that incorporate this Partnering Agreement.	Cost will be managed and the final amounts due will be determined in accordance with the provisions of the bi-party NEC contracts that incorporate this Partnering Agreement.
	Defects will be managed in accordance with the provisions of the bi-party NEC contracts that incorporate this Partnering Agreement.	Defects will be managed in accordance with the provisions of the bi-party NEC contracts that incorporate this Partnering Agreement.
Amount of payment if target is improved upon or achieved	Nil	Nil.

Other Likely Contributions:

Contribution: To represent the interests of the building users in cooperation with other members of the project team .

To represent the Client's financial interests in cooperation with other members of the project team.

To provide structural engineering design skills in cooperation with other members of the project team .

To provide environmental services engineering design skills in cooperation with other members of the project team.

To provide design and construction management skills in cooperation with other members of the project team.

To provide groundworking skills in cooperation with other members of the project team.

To provide structural steelwork design, manufacturing and assembly skills in cooperation with other members of the project team.

To provide external envelope design, manufacturing and assembly skills in cooperation with other members of the project team.

Schedule of Core Group Members

Date of last revision: 25 January 2001

The Core Group members are the *Client* and the following:

Name of Partner	Management Consultants plc	Building Design Associates
Address and Contact Details	Management House, Planning Way, London EA8 7PM, Telephone: 020 7765 5016, Fax: 020 7765 5016	Architect House, Design Way, London EA8 8AE, Telephone: 020 7890 2016, Fax: 020 7890 2216
	E-mail dougjones@managementcons.com	E-mail marysmith@buildingdesign.com
Joining date:	8 January 2001	8 January 2001
Leaving date:	1 March 2002	1 March 2002

Other Likely Core Group Members:

- The building users' representative.

- The Client's financial representative.

- The structural engineering design organization.

- The environmental services engineering design organization.

- The contractor providing design and construction management.

- Those specialist contractors likely to make key contributions.

Setting up the Own Contracts

Contracts will then be set up with each Partner when the Baseline Document has been drafted providing sufficient definition for the selected contract strategy. In most cases the *Client* will also be a Party to the contract (as the *Employer* in terms of ECC, PSC or ECSC contracts). It is quite conceivable that a specialist contractor be brought into the Partnering team through a subcontract to the Contractor. In this case the *Client* is not a Party to the (sub) contract, but the *Client's* objectives are still served by the use of the same Option Data for Option X12.

The methodology for ECC and PSC

In the first statement of Contract Data provided by the *Employer* for each contract between the *Client* and another Partner, Option X12 is stated as being one of the options included in the *conditions of contract*. For example: In an ECC 2nd edition contract under 1 General:

- The *conditions of contract* are the core clauses and the clauses for Options C, H, K, N, P, R, Z and X12 (contained in the NEC partnering option, Option X12, June 2001 published by the ICE) of the 2nd Edition (November 1995) of the NEC Engineering and Construction Contract.

Depending on the final form of future editions, which are likely to include X12 within the available secondary option clauses, this may be revised to

- The *conditions of contract* are the core clauses and the clauses for Options C, X1, X2, X6, X9, X12, Y(UK)1, Y(UK)3 and Z of the   Edition (   ) of the NEC Engineering and Construction Contract.

In the Contract Data provided by the *Employer* under data required for 'Option X12: Partnering', insert the completed Data (*Client* details, *Client*'s objectives and Partnering Information) as provided above.

Until future editions are available which include Option X12; Partnering, it may be advisable to include this additional statement:

- The Option X12 Conditions are {either download the conditions from www.thomastelford.com and included here or state where they are located.}

The methodology when using an ECSC

Insert in the last Contract Data statement, the following:

The *conditions of contract* are the first edition (July 1999) of the NEC Engineering and Construction Short Contract and the following additional conditions:

Option X12 Conditions, (either download the conditions from www.thomastelford.com and insert here or state where they are located, e. g. contained within the NEC partnering option, Option X12, June 2001 published by the ICE), for which the following data is required:

Insert here the data in the same format as given above for ECC and PSC, namely:

The *Client* is

The *Client*'s objective is

The Partnering Information is in document referenced

Other common data as agreed in the Project Strategic Plan will also be inserted into the respective Contract Data statements in each Partner's contract.

Conclusion

The above example is provided as a guide only and is not claimed to be complete. Use of the NEC Partnering Option will soon develop its own experience base and doubtless overtake some of the methodologies given in this Chapter. The Authors would welcome feedback for use in future editions of this Guide and may be contacted through the publishers, Thomas Telford Limited.

Appendix

● Full Text and Guidance Notes

- **a new**
- **engineering contract**
- **document**

JUNE 2001

- # the NEC
- # partnering
- # option

OPTION X12

Published for the Institution of Civil Engineers by Thomas Telford
Publishing, Thomas Telford Limited, 1 Heron Quay, London E14 4JD.

The New Engineering Contract is a family of standard contracts, each of
which has these characteristics:

- Its use stimulates good management of the relationship between the
  two parties to the contract and, hence, of the work included in the
  contract.

- It can be used in a wide variety of commercial situations, for a wide
  variety of types of work and in any location.

- It is a clear and simple document - using language and a structure
  which are straightforward and easily understood.

The NEC Partnering Option is an option that can be used in all NEC contracts except
the Adjudicator's Contract. The other documents in the family are as follows:

The Engineering and Construction Contract (ECC)
The Engineering and Construction Contract – Options A-F
- Option A: Priced contract with activity schedule
- Option B: Priced contract with bill of quantities
- Option C: Target contract with activity schedule
- Option D: Target contract with bill of quantities
- Option E: Cost reimbursable contract
- Option F: Management contract
The Engineering and Construction Subcontract
Guidance notes for the Engineering and Construction Contract
Flow charts for the Engineering and Construction Contract
The Professional Services Contract
Guidance notes and flow charts for the Professional Services Contract
The Engineering and Construction Short Contract
The Engineering and Construction Short Subcontract
Guidance notes and flow charts for the Engineering and Construction Short Contract
The Adjudicator's Contract
Guidance notes and flow charts for the Adjudicator's Contract

The NEC Partnering Option: First Edition, June 2001

9 8 7 6 5 4 3 2 1

ISBN   0 7277 2976 4

The NEC Partnering Option

Guidance Notes

Introduction

A partnering contract, between two Parties only, is achieved by using a standard NEC contract. This Option X12, which puts the NEC Partnering Option into a contract, is used for partnering between more than two parties working on the same project or programme of projects. The Partnering Option is used as a secondary option common to the contracts which each party has with the body which is paying for its work. The parties who have this option included in their contracts are all the bodies who are intended to make up the project partnering team. The Partnering Option does not create a multi-party contract.

The option does not duplicate provisions of the appropriate existing conditions of contract in the NEC family that will be used for the individual contracts. It follows normal NEC structure for an option, in that it is made up of clauses, data and information.

The content is derived from the Guide to Project Team Partnering published by the Construction Industry Council (CIC). The requirements of the CIC document that are not already in the NEC bi-party contracts are covered in this option. The structure of the NEC family of contracts means that this Partnering Option will not work unless an NEC contract is used.

The purpose of the Option is to establish the NEC family as an effective contract basis for multi-party partnering. As with all NEC documents, it is intended that the range of application of this document should be wide. By linking this document to appropriate bi-party contracts, it is intended that the NEC can be used

- for partnering for any number of projects (i.e. single project or multi-project),
- internationally,
- for projects of any technical composition, and
- as far down the supply chain as required.

The option is given legal effect by including it in the appropriate bi-party contract by means of an additional clause (Option X12 in the ECC for example). This option is not a free standing contract but a part of each bi-party contract that is common to all contracts in a project team.

The underlying bi-party NEC contract will be for a contribution of any type, as contractor or consultant for example, the work content or objective of which is sufficiently defined to permit a conventional NEC contract to be signed. The content need not be well defined for a cost reimbursable or time based contract in the early stages.

Parties must recognise that by entering into a contract which includes Option X12 they will be undertaking responsibilities additional to those in the basic NEC contract.

A dispute (or difference) between Partners who do not have a contract between themselves is resolved by the Core Group. This is the Group that manages the conduct of the Partners in accordance with the Partnering Information. If the Core Group is unable to resolve the issue, then it is resolved under the procedure of the Partners' Own Contracts, either directly or indirectly with the *Client,* who will always be involved at some stage in the contractual chain. The *Client* may seek to have the issues on all contracts dealt with simultaneously.

The Partnering Option does not include direct remedies between non-contracting Partners to recover losses suffered by one of them caused by a failure of the other. These remedies remain available in each Partner's Own Contract, but their existence will encourage the parties to compromise any differences that arise.

This applies at all levels of the supply chain, as a Contractor/Consultant who is a Partner retains the responsibility for actions of a subcontractor/subconsultant who is a Partner.

The final sanction against any Partner who fails to act as stated in the Partnering Option is for the Partner who employed them not to invite them to partner again

There are many scenarios in setting up a project. The NEC family of contracts with the Partnering Option is sufficiently flexible to deal with them. For example, the contract may be a Professional Services Contract with a Contractor for a feasibility study. Subsequently that Contractor may do later work using the Engineering and Construction Contract. The Contractor may be a Partner during both stages of his contribution to the project.

Additional contract data for the Option

The *Client* is the Party for whom the projects are being carried out. He may also be the Employer in an NEC contract.

The *Client*'s objective is the objective for the 'programme of projects' if more than one or for 'the project' if only one. The objective should be expressed quantitatively if possible (the business case). It should also include the partnering objectives.

Partnering Information includes any requirements for

- use of common information systems, sharing of offices,
- attendance at Partners' and Core Group meetings,
- participation in partnering workshops,
- arrangements for joint design development,
- value engineering and value management,
- risk management, and
- other matters that the Core Group manages.

This information should not duplicate requirements in the bi-party contracts.

The additional contract data for the Option, like other Contract Data in the NEC contracts, does not change. The Schedule of Partners and the Schedule of Core Group Members, like the Activity Schedule and other schedules referred to in Contract Data do change from time to time. The following are samples of the typical information required in these schedules.

Schedule of Partners

Date of last revision: ....

The Partners are the following.

Name of Partner	Representative's Address and contact details	Contribution and objective	Joining date	Leaving date	Key Performance Indicator	Target	Measurement arrangement	Amount of Payment if the target is improved upon or achieved *

*Enter *nil* in the last column if there is to be no money incentive

Schedule of Core Group Members

Date of last revision: ....

The Core Group members are the *Client* and the following.

Name of Partner	Address and contact details	Joining date	Leaving date

Including the option in the Own Contracts

This option is incorporated into the Own Contract of a Partner as follows.

1. Add "X12 (published by the ICE June 2001)" to the list of Options set out in the first entry in Part one of the Contract Data.

2. Add the following entry to the Contract Data in each bi-party contract:

"Option X12

The *Client* is ....

....

    Address...

....

....

    Contact Details...

....

    Telephone................................... Fax...

.... E-mail...

The *Client*'s objective is ....

....

....

....

....

....

The Partnering Information is in ....

....

....

....

GUIDANCE NOTES ON CLAUSES

Identified and defined terms

Clause X12.2 (1)
The point at which someone becomes a Partner is when his Own Contract (which includes the Partnering Option) comes into existence. They should then be named in the Schedule of Partners, and their representative identified.

Clause X12.2 (3)
Not every Partner is a member of the Core Group.

Clause X12.2 (5)
There are two options for subcontractor partners. Either the amount payable cascades down if the schedule allocates the same bonus/cost to the main contractor and subcontractor, or the main contractor absorbs the bonus/cost and does not pass it on.

Working together

Clause X12.3 (5)
The Core Group organises and holds meetings. It produces and distributes records of each meeting which include agreed actions. Instructions from the Core Group are issued in accordance with the Partner's Own Contract. The Core Group may invite other Partners or people to a meeting of the Core Group.

Clause X12.3(8)
The Partners should give advice and assistance when asked, and in addition whenever they identify something that would be helpful to another Partner.

Clause X12.3 (9)
A subcontractor/subconsultant may be a Partner, but the general policy on this should be decided at the beginning of the Project. The Core Group should advise the Contractor/Consultant at the outset if a subcontractor/subconsultant is to be asked to be a Partner. A subcontractor/subconsultant who the Core group decides should be a Partner should not be appointed if he is unwilling to be a Partner.

Incentives

Clause X12.4 (1) (also "X12.1 (1) and X12.3(3)")
If one partner lets the others down for a particular target by poor performance, then all lose their bonus for that target. If the Employer tries to prevent a target being met, he is in breach of clause 10.1.

There can be more than one KPI for each partner. KPIs may apply to one Partner, to several partners or to all partners.

An example of a KPI

KPI	Number of days to complete each floor of the building framework
Target	14 days
Measurement	Number of days between removal of falsework from the entire slab and from the slab below
Amount	Main contractor - £5,000 each floor
	Formwork and concrete sub-contractor - £2,000 each floor
	Structural designer - £750 each floor

Clause X12.4 (2)

The *Client* should consult with the other Partners before adding a KPI. The effect on subcontracted work should be noted; adding a KPI to work which is subcontracted can involve a change to the KPI for a subcontractor/subconsultant.

Option X12: Partnering

Actions	**X12.1**	(1) Each Partner works with the other Partners to achieve the *Client*'s objective stated in the Contract Data and the objectives of every other Partner stated in the Schedule of Partners.

(2) Each Partner nominates a representative to act for it in dealings with other Partners.

(3) The Core Group acts and takes decisions on behalf of the Partners on those matters stated in the Partnering Information.

(4) The Partners select the members of the Core Group. The Core Group decides how they will work and decides the dates when each member joins and leaves the Core Group. The *Client*'s representative leads the Core Group unless stated otherwise in the Partnering Information.

(5) The Core Group keeps the Schedule of Core Group Members and the Schedule of Partners up to date and issues copies of them to the Partners each time either is revised.

(6) This option does not create a legal partnership between Partners who are not one of the Parties in this contract.

Identified and defined terms **X12.2** (1) The Partners are those named in the Schedule of Partners. The *Client* is a Partner.

(2) An Own Contract is a contract between two Partners which includes this option.

(3) The Core Group comprises the Partners listed in the Schedule of Core Group Members.

(4) Partnering Information is information which specifies how the Partners work together and is either in the documents which the Contract Data states it is in or in an instruction given in accordance with the contract.

(5) A Key Performance Indicator is an aspect of performance for which a target is stated in the Schedule of Partners.

Working together **X12.3** (1) The Partners work together as stated in the Partnering Information and in a spirit of mutual trust and cooperation.

(2) A Partner may ask another Partner to provide information that he needs to carry out the work in his Own Contract and the other Partner provides it.

(3) Each Partner gives an early warning to the other Partners when he becomes aware of any matter that could affect the achievement of another Partner's objectives stated in the Schedule of Partners .

(4) The Partners use common information systems as set out in the Partnering Information.

(5) A Partner implements a decision of the Core Group by issuing instructions in accordance with its Own Contracts.

(6) The Core Group may give an instruction to the Partners to change the Partnering Information. Each such change to the Partnering Information is a compensation event which may lead to reduced Prices.

(7) The Core Group prepares and maintains a timetable showing the proposed timing of the contributions of the Partners. The Core Group issues a copy of the timetable to the Partners each time it is revised. A Partner incorporates information in the timetable into its Own Contract programme.

(8) A Partner gives advice, information and opinion to the Core Group and to other Partners when asked to do so by the Core Group. This advice, information and opinion relates to work that the other Partner is carrying out under its Own Contract and is given fully, openly and objectively. The Partners show contingency and risk allowances in information about costs, prices and timing for future work.

(9) A Partner notifies the Core Group before subcontracting any work. A Partner is responsible under its Own Contract for the actions and inactions of its subcontractor.

Incentives X12.4 (1) A Partner is paid the amount stated in the Schedule of Partners if the target stated for a Key Performance Indicator is improved upon or achieved. Payment of the amount is due when the target has been improved upon or achieved and is made as part of the amount due in the Partner's Own Contract.

(2) The *Client* may add a Key Performance Indicator or associated payment but may not delete or reduce a payment stated in the Schedule of Partners.

ACKNOWLEDGEMENTS

The First Edition of the NEC Partnering Option was produced by the Institution of Civil Engineers through its New Engineering Contract Panel. The members of the Partnering Option Working Group are:

Peter Higgins, (Chairman) BSc, CEng, FICE, FCIArb
Dr. Martin Barnes, BSc(Eng), PhD, FREng, FICE, FCIOB, CIMgt, ACIArb, MBCS, FRSA, FInstCES, FAPM
Ms Frances Forward, BA(Hons), DipArch, MSc(Const Law), RIBA, FCIArb
A. J. McNaughton, BEng, MSc, CEng, MICE

The original New Engineering Contract was designed and drafted by Dr Martin Barnes then of Coopers and Lybrand with the assistance of Professor J. G. Perry of The University of Birmingham, T. W. Weddell then of Travers Morgan Management, T. H. Nicholson, Consultant to the Institution of Civil Engineers, A. Norman then of the University of Manchester Institute of Science and Technology and P. A. Baird, then Corporate Contracts Consultant, Eskom, South Africa.

The members of the New Engineering Contract Panel are:

P. Higgins, (Chairman) BSc, CEng, FICE, FCIArb.
P. A. Baird, BSc, CEng, FICE, M(SA)ICE, MAPM
Dr. M. Barnes, BSc(Eng), PhD, FREng, FICE, FCIOB, CIMgt, ACIArb, MBCS, FRSA, FInstCES, FAPM
A. J. M. Blackler, BA, LLB(Cantab), MCIArb
P. T. Cousins, BEng(Tech), CEng, MICE
L. T. Eames, BSc FRICS, FCIOB
Ms F Forward, BA(Hons), DipArch, MSc(Const Law), RIBA, FCIArb
Professor J. G. Perry, MEng, PhD, CEng, FICE, MAPM
N. C. Shaw, FCIPS, CEng, MIMechE
T. W. Weddell, BSc, CEng, DIC, FICE, FIStructE, ACIArb
F. N. Vernon, BSc, CEng, MICE (Technical Adviser)

Printed and bound in Great Britain